CRITICAL ISSUES IN BIOMEDICAL SCIENCE

A GUIDE FOR BIOCHEMISTRY AND MOLECULAR & CELL BIOLOGY GRADUATE STUDENTS, POSTDOCTORAL FELLOWS, AND JUNIOR FACULTY

By
Leland L. Smith
Professor Emeritus
Department of Human Biological Chemistry and Genetics
University of Texas Medical Branch, Galveston, TX

ISBN 0-7414-1234-9

Published by:

PUBLISHING.COM

519 West Lancaster Avenue
Haverford, PA 19041-1413
Info@buybooksontheweb.com
www.buybooksontheweb.com
Toll-free (877) BUY BOOK
Local Phone (610) 520-2500
Fax (610) 519-0261

Printed in the United States of America

Printed on Recycled Paper

Published December-2002

Dedication

This screed is dedicated to those
who may be in need.

CONTENTS

CONTENTS

CONTENTS

CHAPTER 1. INTRODUCTION

And he that increaseth knowledge increaseth sorrow
- Ecclesiastes 1,18

As we pass the millennium, at the end of a century filled with immense advances in science, technology, engineering, and medicine, we stand on the verge of yet further vast strides in the understanding of Nature. Through the miracle of modern molecular biology and associated sciences we shall see further control over health matters, over disease. Shortly, perhaps we shall gain understanding of perception, of memory. The memories of great scholars and scientists may be tapped and recorded like the present transfer of data from one computer to another. We shall see the total synthesis of living forms from abiogenic materials. There is no limit.

The miraculous modern advances in medical practice are a direct product of generous research funding by our federal government. Modern biomedical research funded by the National Institutes of Health (NIH) and National Science Foundation (NSF) has made very substantial improvements in human health since the end of World War II. Advances against human ailments of all sorts are the case; note particularly the sharp decline in Acquired Immune Deficiency Syndrome (AIDS) deaths since 1995, resulting from very generous NIH funding. We may reasonably expect these advances to continue to improve human health in the foreseeable future, given continued national prosperity and government financial support.

However, as with any domestic endeavor where generous federal funding is available, there are unfavorable aspects that detract from the many benefits. One recalls the plaintive remark of Jimmy Mahoney (a.k.a. Paul

Ackermann) of Bertold Brecht's Aufstieg und Fall der Stadt Mahagonny: *"Aber etwas fehlt"* (But something is lacking)!

We see erosions of American traditions in our schools and of trust in ever so many institutions. Students are being led into odd and devious ways. The radicalization of our culture is typified by the termination in 1990 of a required first year course "Western Culture" at Leland Stanford University, to be replaced with the course "Culture, Ideas, and Values" focused on racism, sexism, and colonialism, a course without grades devoted to self-esteem. Self-esteem, meaning feeling good, has replaced the older meaningful term self-respect.

The University of Chicago classics tradition is similarly politicized, dumbing down with emphasis on race, "gender", and class to retain student enrollment. Companion revisions of our national history are exemplified by the National History Standards developed by the University of California at Los Angeles, revisions that dwell on fixations of race, ethnicity, and sex but not on political freedom. The nonce word "diversity" is mentioned repeatedly, liberty not once![1]

In science education a change in basic philosophy has occurred. Where once student research was conceived as a means of teaching experimental approaches to a problem, student research today is a lower paid part of goal-oriented research of the professor, the principal investigator.

The loss of trust in public matters has developed in part because of increased awareness resulting from greatly improved communications. New ideas may be advanced without the thoughtful restraints that once pervaded communications, namely the publication of printed tracts and books, work that took time to arrange. Hypercommunication avoids these constraints.

The ease with which they who control communications media may advance their notions, services, and products poses a burden to civilization sure to cost more money than realized. A monopoly on expression of thoughts potentially affects scholarship of all sorts, and progress in science, to be sure.

We are entering a new age where past standards of conduct are in flux. Hucksters have sold their wares since antiquity; politicians have gained sway over us with their glib assurances of better times. We are inured to the mendacity of politicians, the duplicity of business practices, and the race for riches in some professions. Fortunately engineering practices have not become quite so shady, or bridges and buildings would collapse and airplanes fall from the sky. Science has also remained relatively free from lies and fraud.

We see two blights affecting science, unique to our country, not encountered in the rest of the world. Our generous funding of science attracts nonscholars who nonetheless are clever, ambitious entrepreneurs willing to use the system for personal gain, even unto misconduct and fraud (see Chapter 6). Secondly, our freedom to do and say brings forth true believers seeking government-condoned intrusions of "creation science" and "intelligent design" (p.85, p.118) into classrooms as science.

Another blight on scholarship and academic duties now developing is that of accountability. One must account their efforts periodically, in various annual reports to their organizations and, more importantly, to their research funding sources. Gone are the days when Nobel Prize laureate (1952) Archer J. P. Martin, while a Robert A. Welch Foundation professor at the University of Houston, asked to provide his annual research report, wrote "I was thinking".[2]

The age-old tenure system protecting the scholar and teacher from political abuse is being replaced and scholarship de-emphasized in favor of accountability. Somehow it is no longer possible to operate biomedical institutions without accountability. During past lean years this was never a concern; now that federal funding largesse is so important, accountability must be had.

We hear "Research is a business. Educating graduate students is a business". Thus, we see the intrusion of sharp business practices into our education system under the guise of accountability. Business management systems are being designed to measure productivity of individuals and of departments. All research, teaching, and patient care efforts must be quantified; metrology reigns supreme. We see such buzz terms as "aggregated metric" and "relative value units". For research, the number of grants, the frequency of grants from NIH, the number of grant dollars, the number and prestige of publications, and the number of citations to published papers are among the items to be measured. Productivity may then be determined by some formula that weights each factor to achieve the desired ends.[3]

For years we at the University of Texas Medical Branch (UTMB) heard of the full-time equivalent for assignment of laboratory space, perhaps 200-800 square feet. Today one buys their space with the overhead contributions from their generous federal grants, in much the same way 18th and 19th century British army officers bought their commissions, a major's costing more than a captain's. Moreover, the successful, funded investigator holds a premier position; he buys his academic appointment but now must keep paying for it. In order to aid these arrangements "The Grant Swinger Papers" is available, grace to the Center for Absorption of Federal Funds.[4]

In mission-based management systems there must be a budget for every mission. To which account do you charge your time for each daily activity? How do we bill for toilet time, lunch time, open discussions among colleagues and with students? Who pays your stipend when you are merely thinking! Indeed, can you do such a thing anymore.

But now we are beginning to see shady practices in the conduct of government-funded biomedical research such that some awareness of the problems must be had by those involved in such ventures. Government funding agencies and their agents may serve the matter well, but the nature of bureaucracy and its political influences may be adding to the problem as much as solving it or serving science.

The student and junior investigator of science may have been attracted to science by natural curiosity, but there are those who were recruited. The motto "Join Science and See the World" insidiously seduces the naïve to become assistants to those senior to them already in practice. It has not ever been otherwise, as junior parties have always begun their careers under the tutelage of masters. There is so much to learn, so many things to know, that arrival at the master's status just takes a long time, even where the student is favored by birth, privilege, and money.

So, one must serve their time, emulate the master (the nonce "role model"), and hope for continuation of support in some way for survival as an independent investigator. Those who opt for employment in industry have other problems, possibly becoming merely a highly skilled specialist or technician knowing everything about the company's business, but also being married to the company, with no ready means of changing jobs outside the specialty.

These problems now receive the attention of professional organizations. In 1996 alone, sessions at three national science meetings were held: "Chemistry and the National Agenda: Ethics and Professionalism", 211th American Chemical Society Meeting, March 1996; "The Conduct of Science: Keeping the Faith", Keystone Symposia, May 1996, Keystone,CO; and "Pathological Science", PITTCON (Pittsburgh Conference), March 1996.

It is my goal here to describe a view of the present status of biomedical research and to outline some of the problems that students and junior researchers may face in their entry into modern funded biomedical research. There are four simple situations that attract our attention, four items that each lead to results impacting biomedical research successes:

(1) Despite generous research funding there still is inadequate funding to support the present system and all who seek to follow a biomedical research career. The limited resources available thus force the independent investigator to

spend inordinate amounts of time and effort merely seeking funds. In the intense competition for funds it appears in some cases that more time is used seeking funds than in conducting funded research.

RESULT: Biomedical research faculty have less or even no time for teaching and for other traditional academic duties.

(2) There are strong recruitment programs aimed at increasing the number of students entering research programs at medical schools and universities. The demand for assistants to perform laboratory work that supports renewed research funding requires more entries than available funding can support.

RESULT: Biomedical research graduate students may be held inordinate lengths of time to complete their graduate degrees merely to support their mentor's research program. Too many such students become "hands" and do not attain independent research careers.

(3) Present federal government research funding arrangements have produced a dependent class of faculty whose allegiance is to their funding agency and not to their academic institution.

RESULT: Funded investigators refuse unwanted academic duties and teaching and flaunt their successful research funding that accords them all sorts of privileges not available to their unfunded peers. Administrations accede to these arrangements because of the sharing of government largesse.

(4) Institutional administrations strongly promote the dependency on generous federal research funding, as the indirect costs and investigators' salaries put on the research grant budget free equivalent funds from university budgets for the administration's own use in other matters.

RESULT: Unfunded faculty may not receive salary increases, adequate office and laboratory space, and other amenities of faculty membership. Tenure may be affected.

These circumstances promote "fundable" research to the

exclusion of fundamental research. Two faculties are built, one with those who teach and conduct scholarly work without generous federal funds, the other with those who have such funding but who do not bother to teach. We hear of two kinds of faculty, productive and nonproductive; productive meaning funded but perhaps less original; nonproductive with less or no funding but perhaps original. Moreover, the independent scientist with innovative ideas is soon isolated with the additional creation of those collaborative "multidisciplinary" research institutes of funded specialists within the university.

This dichotomy may generate an undercurrent of discord that affects morale and efficiency. Costs of achieving the many good results increase dramatically. Built into the system is the expectation that "expense measures quality"! In rare meetings of faculty with our president and dean of medicine I imprudently raised the issue of the creation of two faculties, those with federal funding and those without funding. I was admonished that those who do not like it could leave the university.

Publish or Perish long ago became Publish *and* Perish if one lacks a grant. Mere innovative science research with minimum begged, borrowed, or stolen equipment does not count even if you achieve international regard. Those with genuinely long-range thoughts about serious human health matters shall have to devise short-term fundable projects or fall out of the system.

One must feel lost without generous research funding. Absent a grant, one is no longer a scientist, a man, or person. Recall Dante's inquiry of Virgil *"qual che to sil, od ombra od omo certo "*(Are you a shade or a real man) and Virgil's reply *"Non omo, omo già fui "*(I am not a man; I used to be a man).

It is the creation of a totally dependent class of investigators that has brought the vast progress in biomedical research but also the present troublesome state of research endeavors that so much affects too many. One recalls the adage: "Whom the Gods would destroy, they first subsidize"! Contrast these circumstances with what has regularly gone before - the lack of funds but the opportunity to think and to conduct important research nonetheless. We recall Ernest Rutherford's remark: "We have no money, therefore we must think".

There was a time when we got a grant to do research; now we do research to get a grant. There was a time when a scientist's remarks were considered either true or false; now, like remarks of politicians and business persons, one must ask "Why is he telling me this?".

With awareness of these bitter remarks, it is possible for the student and junior investigator to enter the profession free of the unpleasant surprises that may occur in the absence of such awareness. What may appear here to be a spirit of pessimism is in fact an expression of optimism, that we can accommodate these problems even if we cannot now overcome them.

One other generalization is in order. As with modern computer science, in modern biomedical research the very rapid advances in methodologies, instrumentation, and protocols tend to make obsolete those who entered the field just short years ago. It is essential to maintain effective growth, a growth made available by knowledge of the vast biomedical literature that we now face. Not only do printed journals inundate the individual investigator, but there are computer network resources available well in advance of release of printed issues.

Short of satisfactory awareness of the literature and of success in research funding the independent investigator risks abject failure. In such event all that is left is to join funded entrepreneurial research operations, leave those competitive appointments for industry or college teaching, or undertake wholly different employment.

No one now builds their own mass spectrometer as was the case observed in my early post-doctoral experiences. Modern instrumentation does not have analog dials and knobs; rather costly computer controlled operations are *de rigueur*. Of course, modern equipment is far advanced over the older now obsolete apparatus of just a few years ago, as also is the science that depends on such improved instrumentation. Now the skills of devising or fashioning one's own equipment are no longer needed. Rather, it be the skills to raise money to buy the advanced equipment that is now required.

Moreover, the rapid advances in methodologies and instrumentation leads to another feature, that of disposables. Not only are reagents bought for one-time use, but disposable lab wares are essential, because they are convenient and do not need costly dishwashing services. Both items allow more rapid progress for competitive renewal of the research proposal. However, the concept of disposable laboratory items also leads to disposable laboratory personnel. He, she, or it who does not shape up, who dreams too much, or does not take direction seriously enough can be left out of the next grant renewal.

Among the more severe problems attendant upon generous financial support of science is the recruitment into science careers of candidates who might do better in other directions. The other unavoidable result is that too many are drawn to science at the Ph.D. degree level, saturating job opportunities. The overly active recruitment into science and engineering that followed the Soviet Union's October 1957 launch of Sputnik brought many into science who now compete with one another for federal research funds, with far more effort being exercised in obtaining money than in the thought processes that advance new science. The immense advances of recent years in molecular and cell biology and in genetics that improve human health have resulted in a similar intense increase in graduate students seeking to enter these fields. Indeed, there has surfaced a notion that the NIH should establish a "molecular" graduate school focused on bioinformatics, genomics, and clinical research. Fundamental research appears not to be included.[5]

The number of Ph.D. degrees awarded in the United States increased from 8,773 in 1958 immediately post-Sputnik to 41,610 in 1995, thus creating a body of educated persons larger than that of suitable employment. However drawn to science, whether by Sputnik or molecular biology successes, the graduate student and junior investigator must become acquainted with certain impediments to research successes, and it is my hope that the present effort poses sufficient stimulus to think critically about these aspects of science not readily obtained in indoctrination classes or textbooks.

I describe here a graduate course I and my faculty colleague David A. Konkel conducted at UTMB for years, until my retirement in 1996. Our course, required of our graduate students in biochemistry, molecular and cell biology, and genetics in the Department of Human Biological Chemistry and Genetics (HBCG), was taken after completion of required standard first year courses. The course was designed using relevant published biomedical literature to awaken the students to critical thinking, to experimental design and critical assessment of papers, and to other important features of modern competitive biomedical research.

General guides to graduate school science studies have appeared,[6] as have specific courses given at several universities, courses that seem to be more and more necessary given the degree to which irregularities occur in biomedical research in the present day. In keeping with a 1994 National Academy of Sciences "Convocation on Science Conduct", courses with names such as "Ethics in Science" and "Fraud in the Chemical Sciences" have been instituted at Florida State University, University of California at San Francisco, University of Iowa, University of Maryland, University of Pittsburgh, and University of Tennessee, *inter alia*.[7] Indeed, instruction in these matters is currently required of all those on federally funded training grants, instruction soon to be required of all research personnel!

Critical guidance into some of these problems must be

more broadly available to students and junior investigators, thus this screed. We chose an approach in which colleague David A. Konkel, thoroughly versed in the latest molecular biology literature, would lead discussions into problems of the topic and I, as senior biochemistry professor, would add comments dealing with chemistry, biochemistry, and the physical sciences, it still being acknowledged that chemistry be "the foundation of all medical and biological science".[8] The balance struck was the same as positive/negative, left/right, Yin/Yang, and all the other bipolar terms which if equitably presented give the student a balance not available in indoctrination course work.

The need for our course was evident years ago, when a student in all innocence gave a seminar based on a seriously flawed journal article and was embarrassed by the error, called to her attention in seminar by someone. Even our best students tended to believe it must be right if it was in print, particularly if it was in one of the magic five journals *Nature*, *Science*, *Proceedings of the National Academy of Sciences USA*, *Journal of Biological Chemistry*, or *Cell*. These five journals constitute the five books of Moses so far as modern molecular biology is concerned. Anything published in these journals is to be believed verbatim, no questions asked. Many of our biology faculty swore by these journals, taught as established fact whatever dogma was just out, and inculcated what I viewed as an uncritical approach to science in our students.

It must be recognized that some build their science careers on having their papers published in these journals, particularly in *Cell* or *Nature*. On the occasion of review for promotion of a favored associate professor to professor the issue arose about his publications in *Cell*. I had asked whether anyone had read any of these papers for content. The answer was no, and to my face I was told that we did not have to read these papers to know they were good ones!

However, John Maddox, sometime editor of *Nature,* has conceded "everything *Nature* publishes will be proved wrong within a measurable period of time". The polymerase chain reaction article by Kary Mullis, 1993 Nobel Prize laureate, submitted to *Nature* was refused publication by Maddox.[9]

One must note that papers in the *Journal of Biological Chemistry, Cell,* and *Proceedings of the National Academy of Sciences USA* are officially advertisements. The costs of journal production are met in part by required page charges in the same manner as paid advertisements in magazines and newspapers. By government fiat 18 USC §1734, such science papers must be labeled "advertisement". Other journals such as those of the American Chemical Society have escaped such labeling only after court actions. It is a matter of some amusement to see the expressions of molecular biology authors when asked about their latest advertisement just published.

We proceeded to demonstrate in our course that papers in respected journals could still be seriously flawed and in need of rejection or revision. Necessarily we picked papers that were flawed, just to establish the simple principle that respected journal origins did not guarantee quality. Our students confronted with the evidence could no longer accept an *obiter dictum* of a true believer on some questioned point, or the glowing comments of those of hierarchical or autocratic command status.

It was our design to reveal error in a dramatic manner, thereby to be remembered better by shocked students. A spade was called a spade. We not only opened eyes but also shattered illusions. Our approach caused consternation among some students, as they had not been previously exposed to excursions into reality. They had not experienced harsh criticism of flawed literature, of genuine error, including their own performance in evaluation of assigned papers.

I had witnessed rather heated arguments in American Chemical Society national meetings long ago when a graduate student, events I still fondly recall clearly. The dearth of similar arguments over key issues in national and international meetings now seems more the experience. This may be result of the "self-esteem" concepts of the current Political Correctness (PC) fad, but also may just reflect that argumentative points no longer are expressed in these meetings, thereby avoiding embarrassment or controversy. Senior scientists may not even be in attendance, as they are likely to attend the more attractive advanced specialty meetings abroad, while sending their junior associates or graduate students, fully unprepared for heated argument, to present material at the large meetings for experience.

Students rarely forgot lessons learned at expense, but some paid a high price. Although we tried not to embarrass students, we did call them WRONG where they were wrong. Wrong like the old Major Bowles Amateur Hour gong, without regard to current "self-esteem" fancy. We gave favorable remarks for sound responses in class and also graded class performance and written work to calculate course grades. None of the "No winners, no losers" business. We know there must be winners and losers in the modern professional ball game business (they must win, you see) and in the federal research funding game but unnecessary elsewhere, wherever PC can be imposed. A few students never regained their composure.

It does seem foolish to expect to overcome competitiveness developed over millions of years of survival of the fittest, or perhaps more appropriately, survival of the survivors, with late twentieth century notions of self-esteem. Besides, Johann Wolfgang von Goethe has provided us in his Coptic Song with guidance for living a prudent life: *"Töricht auf Bessrung der Toren zu harren! Kinder der Klugheit, O habet die Narren Eben zum Narren auch, wie sichs gehört!"*. (Foolish to seek improvement in dunces! Children of wisdom, let dunces be dunces, as it should be!).

In that our course was based on reading published journal papers right there for posterity, we had to address which sorts of flawed papers to assign. Clearly, opportunity for bias exists in the selection of material. Moreover, some questioned whether good teaching result from use of flawed materials, or whether use of sound papers with detailed analysis of the good points be a better approach. Which lesson would you rather learn of science publications: error exists and must be recognized lest more error follow, or, good work exists and one should do good work. The only issue is that if no one tells you, how do you know good from bad? Does it matter?

Thus our selection process was devised. There are ever so many sound papers in the biomedical science literature, indeed in literature of any subject, papers that are sound though generally unread and uncited by other researchers. Were scholars to follow their curiosity and interests, this result is about what might be anticipated - that the literature would fill up with scholarly papers about subjects not of broad interest to large bodies of readers.

Another sort of result occurs with those selling their services to industry, government, not-for-profit foundations, or the grants-dependency process - the goal-oriented research project, potentially devoid of intellectual curiosity but feasible and fundable. Necessarily, another literature is built of papers with goals different from the mere disclosure of works of love by scholars. We see the creation of new means of self-advancement, of advertisement if you please. Motives for publishing work for promotion, to build a reputation, and to secure the next project funding are revealed. Where even remote opportunities for commercial exploitation are realized, patent protections are sought. A different literature emerges, one rich in problems, error, and worse.

Moreover, we see goal-oriented literature described as "basic research" by its proponents. There was once "pure" or "basic" research and "applied" research or development, but these terms have gone out of style. The once readily distinguished terms are described by Glenn T. Seaborg: "The distinction between basic and applied research lies in the motivation behind the research, and the criteria applied to determine what work shall be undertaken, and what changes shall be made in the lines of investigation as the study develops. In basic research, the motivating force is not the attainment of utilitarian goals, but it is a search for a deeper understanding of the universe and of living phenomena, whereas in its keynote is intellectual curiosity".

Today, funded projects necessarily become "basic", thereby gaining unearned credit. Unfunded victims of politics or budget restraints fare differently. We have the Hubble space telescope in orbit after some difficulties,[10] but the creative notion to drill the Mohole into the Mohorovicic discontinuity between the earth's crust and mantle is no more, victim of high costs and technical difficulties. Also, the supercollider particle accelerator designed to examine subatomic structure is no more, again victim of finances and choice made in favor of the manned space station for military and business use. Choices must be made.

We are already inured to misuse of "science" in social science, "creation science", and "scientific wrestling", but now we must abuse the words "research" (meaning funded research) and "basic" as well. Thus is gained unearned credit.

Necessarily, our course involved arbitrary selection of material to be discussed, and we make no apology for our selections. Others formulating similar courses may use their own choices, not ours. The papers we chose to use were carefully selected to meet several of the following criteria: (1) With some relation, however extended, to biomedical research; (2) By prominent investigators; (3) From prominent laboratories or universities; (4) In respected peer-reviewed journals; (5) Both within and without the experience of the students; (6) With debatable points for teaching, without closure; (7) With recognizable minor/editorial blemishes; (8) With error, detectable or no at the time; (9) With major flaws mitigating acceptance of the work; and (10) Items disclosing research, publishing, funding, and related fashions.

From year to year we varied some of the papers used in class, thereby to expose our students to newer cases where feasible, to experiment with other aspects of critical reading, but in the case of fraudulent items to ensure that remarks from classmates from prior years did not give away our game plan. For each paper selected for use, for each point made in class, there are many other papers that could have been chosen.

Those creating a similar course for these purposes will necessarily use papers from their own specialty fields for classroom use. The basic concept of enlightened revelation of error can be applied in all other sciences as well. Our attention to biomedical sciences, particularly molecular biology, originated from students' needs together with the immensely rich selection of published papers now available for instruction.

In this matter our selection of papers included those from premier laboratories, universities, and journals, thereby to tempt students to give undue credit to flawed work, even to disregard flaws in work from Harvard University Medical School. This ploy generally failed, as our students were usually unacquainted with the Harvard Syndrome, and gave little weight to renowned laboratories. Another worldwide premier university, such as the Eidgenössischen Technischen Hochschule (ETH), Zürich, was totally unknown to our students.

This insensitivity to Harvard University is commendable in the naïvité of youthful investigators, but there eventually develops an "if it's from Harvard (or NIH) it has to be good" notion that simply is insupportable by evidence. The more prescient mind escapes such conceit, as did Bertrand Russell of Cambridge University: "Against my will, in the course of my travels, the belief that everything worth knowing was known at Cambridge gradually wore off".[11]

Our selection of flawed papers necessarily included items from the five journals of special interest *Cell, Journal of Biological Chemistry, Nature, Proceedings of the National Academy of Sciences USA,* and *Science* and from three premier chemistry journals *Journal of the American Chemical Society, Journal of Organic Chemistry,* and *Angewandte Chemie International Edition* but also from

Advances in Cancer Research, Biochemical and Biophysical Research Communications, Biochimica et Biophysica Acta, Cancer, Discussions of the Faraday Society, FEBS Letters, Journal of Membrane Biology, Journal of Molecular Biology, Life Sciences, Molecular and Cell Biology, Nucleic Acid Research, and *Proceedings of the Society for Experimental Biology and Medicine*; thus careering through a gamut of journals of divers properties.

The several published papers we criticized were from prominent laboratories of premier universities such as Harvard University, Harvard University Medical School, Columbia University, Cornell University, Eidgenössischen Technischen Hochschule, University of California Berkeley, University of Southern California, Massachusetts Institute of Technology, National Institutes of Health, National Bureau of Standards, Bell Telephone Laboratories, and USSR Academy of Sciences, but also from several less well known sources.

Notice that many of the papers we selected were at the forefront of developing science at the time of their publication. These papers reveal a characteristic pattern unique to early innovation, quite different from the pattern of papers appearing in later, better informed development periods. Absent adequate background information or in naïve unawareness of such information there is greater opportunity for misunderstanding and error in emerging science. In this matter, early investigators in innocent error and our students and junior colleagues in naïvité are on the same level when addressing the question of whether a given remark, datum, or conclusion be sound. We see freedom of thought but with a similar chance for error among very young grade-school students of science before informed thinking is available to them.

Once early innovative discoveries attract interest, the increased development of a subject reduces the likelihood of innocent error, as there is then available more background information, more generalized principles, ideas, and methods. There follows an exploitation period of minor ideas and "chink-filling" and a mature orthodox, funded period on the "cutting edge" of science. Standardized thinking a la Walter Lippman: "When all think alike, no one thinks much" may come to predominate.

I personally experienced the growth from early exploratory activities to mature status of two major research topics: steroid hormones and their synthetic analogs from the late 1940s to early 1960s, and oxygen biochemistry from the early 1960s. Both cases led to greatly advanced understanding via biomedical research and to vastly improved medical practice. In both cases patterns of early random excursions into dark areas, fancy, and error were followed by maturation of the topic less subject to fancy and error than to other misadventures. It is from the second of these experiences that oxygen biochemistry was selected as source of papers examined in Chapter 2.

In our discussions of flawed literature we regularly emphasized several key steps. From the very first words of the title of the paper we are alerted to what may be expected thereafter. A title with misspelled or inappropriate words, acronyms, fad or nonce words, PC terminology, and assertive sentences tell us that the authors are ready to play the game, wherever it takes them. We also recognize that the care with which an article is prepared for publication probably reflects the care of the authors in thinking and in conducting their work. In essence authors may announce that they are not scholars, that they do not possess an independent spirit of inquiry, but will use whatever nonce ploy is suitable for advantage.

Once so warned, one must be cautious for worse matters. Where there is poor English (perhaps excepting authors for whom English is not their native language), anticipate poor science. Where there is fad terminology, there may be fad science. Where there is PC, expect other serious error. Nonetheless, one must avoid premature harsh judgement, as good science is now reported in inelegant writing styles. We are reminded of the adage "Don't judge a book by its cover" and must remain, accordingly, between the Scylla and the Charybdis.

It was always a pleasure to come to class, as we were doing things that were needed to be done, and we could see good results directly. The students seemed to enjoy class as well. This is not usually the case in other medical school didactic teaching, particularly of biochemistry, where a healthy dislike is instantly the case for many freshman medical students and, alas, also for our biology graduate students.

This sort of course offered both instructors and students an otherwise unavailable opportunity to see what thought processes each had. Close contact with graduate and medical students at this level cannot occur in didactic lectures, but the intellectual strengths of each student are immediately apparent in our small interactive class. In turn, students can see just what sort of committed efforts it may take to achieve senior faculty status.

On several occasions students spontaneously told us of the value the course had for them, that we had opened their eyes to the reality of the biomedical research world, to their advantage. Somewhat later, several who achieved faculty status told us that our course was their most valuable experience in graduate school. It is so much more rare that medical students ever told me that I had helped them, done the right thing in my teaching.

I offer herein a concise text that should be useful to others designing and implementing their own graduate course work in science areas where this need is recognized. The text here developed not only describes the course as we offered it for years but also provides resources for further exploration of some topics, particularly error and misconduct so much of current interest. I also include comments of personal research and teaching experiences with medical and allied health students suited to the occasion.

In offering personal anecdotes and other remarks I have not attempted to resolve issues so much as to raise them. Faculty using these approaches can introduce topics befitting discussions to whatever length seems appropriate for the class. Moreover, our coverage of these items is not meant to be definitive, as there are now books published on each subject. Neither is our treatment designed to be directly or subliminally judgmental, as is the case with so many articles and books dealing with science misconduct. It is our notion that anyone who really does not understand the role of the scholar among practicing scientists will probably not be helped by required courses, journal articles, or monographs outlining ethical principles.

Certain more provocative remarks have been included from place to place to awaken, or even startle, the reader and to stimulate additional thought processes. It is not my intent to offend or insult but to provoke thought! If I fail in this matter for some, perhaps for others there is a valuable gain.

The specific published articles discussed in Chapter 2 should be understood to be merely examples of the points taken. Others using our guide will surely use different papers of their own choice to illustrate issues. This choice is crucial for the two class sessions dealing with specialty topics, where the expertise of the two instructors provides direction. Obviously if specialty topics are to be used, the specialist must be fully acquainted with the topic at the expert level.

In this endeavor, both I and David A. Konkel examined the latest journal literature available on a daily basis, trying to keep aware of developments pertinent to our own scholarly and teaching interests.

CHAPTER 2. THE COURSE

Do not feel absolutely certain about anything!
- Bertrand Russell

Our course was designed for the second year following required biochemistry, molecular and cell biology, genetics, and introductory laboratory course work of the first year of graduate study. Attempts to truncate the course into the first year (so as to free students for research full-time!) by research oriented faculty requiring student labor to advance their own work failed so long as I was involved, as the timing and teaching of a course must be the prerogative of the instructors, not of those who do not teach.

The course as here described is no longer the same, given my 1996 retirement and a change in direction for my biology colleague into research administration. These two developments ensured instant obsolete status for both of us anent the continuing growth of the biomedical literature now no longer being examined daily by either of us for interest and possible use in teaching. Moreover, the graduate programs at UTMB have been changed, with a first-year core curriculum for all entering students. The nature of the Critical Concepts course shall surely be altered. Others have taken up the course now somewhat different in concept and arrangement.

1. Our Students

Our graduate students were bright and interested but of diverse backgrounds, including foreign students with suitable preparation but with some English deficiencies. There were five types in our HBCG graduate program: (1) Graduate students (M.A., Ph.D. degrees), usually biology majors with relatively poor chemistry backgrounds; (2) Students in the UTMB combined MD/PhD program, seeking both degrees, generally better prepared; (3) Unsuccessful medical school applicants, seeking graduate degrees to improve chances of medical school admission; (4) UTMB employees seeking graduate degrees to garner promotion; and (5) Occasionally a scholar.

With this mix of students and the dumbing down occurring in graduate education, we had to create a course that would include features of chemistry and biochemistry, the basics supporting modern medicine, along with the greater interests of the students in molecular and cell biology.

Our departmental graduate program entrance requirements were essentially the same as for medical school; thus few students had advanced chemistry courses. Students had only one year of English but had several biology courses. Our foreign language requirement had been dropped decades ago, with the usual specious excuse that computer science would take up the void. This did not happen, and our weaker students were excused from one more burden of education.

The lack of general science background of the students surfaced in several classroom discussions. One was my account of the recent synthesis of an enzyme, HIV-1 protease, from D-amino acids. When asked whether such a D-enzyme could be active on enantiomers of the natural L-substrates, our students could not agree that such was mandated by Nature.[1]

Our course was fashioned into two parts, one involving student criticism of published work, the second requiring student compositions for class criticism. Students wrote a brief critique of each assigned paper and used their critique in class, where they were called upon, if necessary by name, to remark on the paper. Was it a good paper? How do you know? Frequently a straw vote on each item's quality was taken. Students were not always correct in their assessments.

Critical discussions followed, often well past the official end of class.

Students were warned not to alter their written critiques during the discussions, and their written work was graded for content but not for English, although errors were marked. The lack of adequate English was ever present, with usual problems of subject-verb agreement (data is), pronoun case, and the like.

Anent split infinitives, my admonition "Split not lest ye be split" ofttimes brought the question "What is a split infinitive anyway?". Always with an apology that their English course was several years ago. Nonetheless, we eschewed teaching remedial English to domestic students and English as a second language for our foreign students. There are many guides to proper English available, also many books describing how to write technical papers.[2]

The common remark "I don't think" always got a YOU MUST THINK! from me, orally in class but ever so heavily on written work. The related "I think not!" now so frequently heard in the public domain likewise announces your incompetence. Recast your sentence; do not announce that you do not think! Let others find out some other way. Deletion of the terminal letter "t" yielding "I think no" would more exactly inform us of your status.

Students were advised to be very careful to avoid error in their writing. One letter exchanged in the word "now" to "not" changes the meaning of a statement. Reliance on computer spell-check programs cannot correct such matters. Also, reliance on computer program charting of data is not free of trouble, as some programs interpolate curvilinear results to linear presentations, and other errors can result. More than a few students used the excuse that the computer did it, and that must be so. However, as for causation, who pushed the keys?

2. Introductory Remarks

We opened the course, our students and the two of us sitting together around a large table, with brief mention of our goals and means: This is a different sort of course, as stipulated in the course description in the University catalog. The course is not intended to impart information, and will essentially be self-taught, with the faculty as guides. Students are expected to participate in class discussions. There shall be no written tests. There shall be no place to hide, no automatic "B" (equivalent to the undergraduate gentleman's "C" grade).

We use Socratic methods; published papers will be assigned to be read and criticized for both good and bad points. The assigned papers may be within but also outside of the students' experience. A written critique of each paper must be prepared before class, the critique to be used in class discussion but handed in for grading at the close of class. Examination of the primary literature may be necessary for writing a proper critique. Criticisms should be those that a contemporary scientist might make. Thus, do not use forward searches of the literature by computer, as such may reveal much more than a contemporary reader could know.

There will be no written tests; grading will be: 15% credit for written critiques, 20% for oral participation, 55% for contents of a written review and related research proposal, and 10% for oral presentation of the review and proposal.

3. Course Outline

The course extended over a 15-week term, meeting for 90 minutes twice a week but often for up to two hours, given the heat of discussion and interest. The course was divided into three formal parts, the first of which, 10-12 class meetings, involved criticism of 25-30 published science papers, as outlined. In Part 1, Critical Discussions of Published Articles, after introductory remarks, assigned journal articles dealing with experimental design, critical assessments, and review papers were covered. Thereafter we had two sessions dealing with special topics: *in vitro* transcription within the experience of most students, and oxygen biochemistry well outside their experience. We took our students outside their experience regularly by design, thereby to provoke thought.

The specialty sessions delved more deeply into problems of the published literature and of necessary later adjustments, not to impart detailed information but to illustrate typical problems in emerging biomedical topics of great interest. It was in this lesson that the expertise of the faculty was paramount, as judgement of each assigned specialty paper required substantial awareness of the literature to the present moment.

There followed assignments of special topics that taught other lessons in modern biomedical science, among which is the "Polywater" episode disclosing the gullibility of scientists but also federal government funding policies of political matters, *vide infra*. The organized part of the course ended with treatment of science fraud generally undetectable from reading the spurious papers.

Students were asked not to use computer bases or to consult with other faculty members or previous students in the course regarding assigned papers. The admonition was usually followed, but in a few cases it was obvious that input from these sources had been used in preparing written criticisms.

In Part 2, Student Compositions, the second part of the course, we exposed the students to how it feels to have their own written papers criticized right there in class by others intent on discovery of deficiency and error. Student papers had to be typed and were graded, with points taken off for poor English in these cases.

Students were asked to prepare a ten-page review of a biomedical topic selected from lists of our special expertise, and thereafter a ten-page research proposal dealing with matters of interest derived from the review. We had 10-20 current molecular and cell biology topics and the same number of topics dealing with steroids, lipids, drug metabolism, oxygen biochemistry, and modern spectroscopy. Students generally chose the biology topics, but each year one or two selected the other topics.

Part 3, Retrospective, the third part of the course (one final session), was our turn to receive criticism from the students about the course and our conduct of it. Over cookies and punch we solicited comments after course grades had been given (We typically gave half the students "A", half "B", with two "C" and two "F" grades over the years), so students might comment knowing their remarks would not influence their grade. Usually only innocuous remarks were elicited about us, as students are wont to avoid future faculty attention to themselves. The course itself generally received good comments about its usefulness in awakening the student to science reality.

Over the term, at first everything went well, but as we unloaded on the students, they became gun shy, began seeing foolishness, split infinitives, gross error, idiocy, and more. Even where fine papers were sneaked in on them, they found deficiencies, mostly trivial. After the initial shock of being called wrong, of having their faith in dogma shattered, of standing on the verge of reality, the students started to read assigned papers critically and to think, which was our "secret" purpose.

The early lessons treated fundamental ideas about experimental science - the proper design of an experiment, assessment of results, and, through review papers, how to come to proper methods and approaches. After these, a more complicated set of problem papers were examined.

4. Experimental Design

The design of an experiment, the plan for study to be conducted, is of crucial importance to most investigations and must be crafted in a manner according some modicum of success. In reality, many initial designs fail but then lead to refined designs that may eventually afford a solution to a problem. The more that is known of a problem and the more closely defined it is, the more easy becomes the design of an experiment to resolve matters. Experience is a great teacher, and the more we have of it, the easier is the approach to new experiments that expand experience and knowledge.

For the most fundamental of issues, that of curiosity, a different set of experiments may be needed than those necessary for exploration of more defined questions. The reflections of youth, the classic naturalist, and the odd scientist yet able to wonder and ponder may generate wholly different notions of experimental design, notions totally unsupported, nay, even excluded by goal-oriented investigators seeking answers limited to stipulated goals, and thus to development rather than original or fundamental research.

Questions such as "I wonder what happens if I do thus and so", "why is grass green", and "could that be so" are quite different from "which disease of the month should I tackle", "how much salt should I put on my dinner", "how long should I stir it", and require very different experimental approaches. There must be an initial definition of a problem, with an adequate search of the literature to acquaint one with what is now known of the matter. Knowledge of prior work is absolutely essential for the effective and scholarly approach. However, "just do what you are told and let me worry about background" prevails in some laboratories.

The significance of the problem and its practicality may be major influences, but curiosity, simple interest, and love of the study should be motivating factors. Generally a hypothesis is posed; one then designs experiments to test the hypothesis. Note that hypothesis-driven research is now *de rigueur* for federally funded research; mere descriptive or discovery work sans hypothesis is not fundable as such, despite the obvious continuous need for new discoveries. One must project possible outcomes of an experiment as they apply to the original problem, yet retain an open mind to alternatives. Let the experimental results tell you the answers to the problem.

There are reliable means of approach in designing experiments. Mention here of a few standard items suggest what may be discussed in detail in class. The method of agreement, that of difference, and the principle of concomitant variation may be discussed as befit. One seeks a single common factor as cause of numerous effects; of several common factors, the one that is different be cause of effect. The intensity of cause is proportional to intensity of effect. A critical experiment may be designed to get a direct answer. Though the direct test is preferred, indirect tests may be all that can be adduced, or a corollary of the hypothesis may be tested: A → B; not B → not A.

Choice of the experimental system is crucial. Model systems may be devised; variables must be defined; assumptions must be clearly recognized as such. The methods to be used must be identified and fully understood. The importance of pilot studies, of inclusion of adequate positive and negative control measures, and of full awareness of the sensitivity of the method must be considered. Increased sensitivity may alter significance. Statistical evaluations should be made.

Perhaps the greatest demand on science is demonstration of the cause-and-effect relationship. Far too many premature and incorrect attributions of cause abound. Putative cause and observed effect may be linked by obvious factors of time and place and by quantitative measures. Absent cause, there must be no effect; more cause gives more effect. Experiments must be reproducible by the investigator and by others in the world.

Endless examples of experimental designs abound in the literature, and any set of papers may be chosen to demonstrate these issues. Examination of any selected example becomes a matter of individual needs adjusted to the maturity of the class and the teaching objectives. In our approach we assigned two papers whose evaluation by our students led to unexpressed but intended lessons. One paper was well within the students' interest, experience, and comprehension; the other was well outside their experience.

With such an experimental design on our part, actually a teaching design, the class became disoriented at the first lesson! How to contend with real criticism, real science, not just classroom didactic lectures.

Both assigned papers were from reputable journals and recognized laboratories. The initial paper was a classic dealing with DNA replication, thus a topic close to each student's heart, with good experimental design making a strong point definitively but with older methods (and with minor flaws of the times). The paper's authors and results were well known to the students, but none had actually seen the original paper. The second paper dealing with environment chemistry of natural bodies of water was far removed from the students' prior experiences but could be understood and criticized with effort. The work involved highly misguided claims that a specific oxidant capable of oxidizing pollutants in water was present in swamp water, thus involving a wholly flawed design ending in fiasco despite complex tabulated data.

Each example will be discussed in some detail so as to establish the nature of our teaching efforts. Those arranging their own critical issues course along these lines may use these or other papers as befit the occasion, the instructors' interests, and the class composition.

The first item is a classic paper that established the conservative replication of prokaryote DNA, a topic of importance at the time. The paper is:

M. Meselson and F. W. Stahl, "The Replication of DNA in *Escherichia coli*", *Proceedings of the National Academy of Science USA*, **44**, 671-682 (1958).

This seminal article showed that complementary strands of DNA did separate in replication, thereby exciting many to study DNA. Although there was an effective experimental design and results provided a definitive conclusion of conservative DNA replication in prokaryotes, there were problems, particularly in attempted extension of results to eukaryote systems.

There are also editorial flaws. On the first line of text (p.671) bacteriophage is misspelled, thus potentially alerting the critical reader of possibly other more serious error. Other minor editorial issues bear mention: "gm" is used for gram instead of "g"; both cesium chloride and $CsCl_2$ are used; the now obsolete N^{14} (^{14}N in modern nomenclature) is used, all unimportant to the thrust of the paper.

Other more important issues occur in an otherwise sound paper. The term "DNA molecule" is used; the DNA dimer is clearly labeled as "original parent molecule", thus raising the question "what is a molecule"? Biology students were confused by such issues; is duplex DNA a molecule or two molecules, a dimer? Secondly, students did not understand that the Gates and Crellin laboratory from which the paper came was a centrifuge laboratory, thus the reason centrifugation and stable isotope ^{15}N were used in the study. Why not use radioactive nitrogen (!) isotopes; just count the samples as usual? Additionally, the major nitrogen isotope ^{14}N is termed "ordinary nitrogen", the ordinariness of the naturally occurring (0.368% atom excess) ^{15}N isotope notwithstanding. None of these thought-provoking items bothered our students, but then this was their first encounter with critical review of published material, and they were not yet ready to read carefully and with critical thinking.

The paper then moves from sound work with *E. coli* DNA into flawed experiments seeking to examine DNA replication in eukaryotic cells. In this phase denaturation of salmon sperm DNA was required, but the greater complexity of the DNA confounded results. The high guanosine/cytosine content of the eukaryotic DNA requires a higher denaturation temperature (107°, boiling water) than does prokaryotic DNA; thus, unbeknownst to the researchers, only partial denaturation of the sperm DNA was had with the methods used. Also, there was a loss of sensitivity in measurement of the ^{15}N content, as eukaryote DNA is much larger than that of *E. coli*.

Unanticipated and imperceptible lessons taken from the class session are several. Even the classic papers mentioned repeatedly in didactic class work may have problems. The most famous investigators make mistakes, are careless in writing, make insupportable extensions from their sound results, and are allowed editorial privilege. Only with other information or experience can sound conclusions be drawn about such issues in original papers, flaws and all.

The second assigned paper exposes another feature of crucial importance, that of faulty design based on misunderstanding of background science. The paper is:

R. G. Zepp, N. L. Wolfe, G. L. Baughman, and R. G. Hollis, "Singlet Oxygen in Natural Waters", *Nature*, **267**, 421-423 (1977).

The possible presence of electronically excited (singlet) molecular oxygen ($^1O_2$) in biological systems was of great interest at the time, and the authors extended this interest to naturally occurring waters. They concluded from extensive data that $^1O_2$ was present in swamp water and that water pollutants would be oxidized by the reactive oxygen species (ROS).

As with the prior paper criticized, minor editorial blemishes are seen early in the paper. There is always question whether editorial flaws be minor or be important, be caused by journal editors or by original authors, or be readily accommodated by readers so as not to interfere with enjoyment and acceptance of the paper. In the present case, the British spelling metre for meter appears in the second paragraph, obviously not the responsibility of the authors. Other flaws include undefined acronym ODS, a split infinitive, use of obsolete mass spectrometry terminology m/e instead of m/z, and confused use of lifetime but also of halflife, half life, halflives, and half lives.

These distracting blemishes are minor compared with the flawed experimental design seeking to show the presence of $^1O_2$ in water samples. 2,5-Dimethylfuran was used to intercept $^1O_2$ in swamp water samples collected for study in washed glass bottles. However, as was frequently the error at the time, the rapidity of interception of $^1O_2$ by the furan was misunderstood for specificity of interception, quite a different concept. The furan intercepts $^1O_2$ very rapidly, and any $^1O_2$ present would be intercepted effectively. However, the furan also reacts with other common oxidants, including hydrogen peroxide (H_2O_2) surely present in aerated water. Elaborate tables of data appear for verisimilitude, all to no avail; a wasted effort to read the paper or examine the data save for the experience of seeing error in a respected journal.

Our students did not know photochemistry or oxygen biochemistry, nor did they check the cited references that would have provided background. They were more concerned with how the collecting bottles were washed than how reliable the use of the interceptor might be. The dimethylfuran oxidation product detected merely evinces the presence of an oxidant in the waters, not necessarily $^1O_2$. It was well known at the time that this was so, that aerated water contains H_2O_2 and that H_2O_2 oxidizes furan derivatives. Subsequent studies specifically address the problem and demonstrate formation of expected H_2O_2.[3]

Neither the authors nor our molecular biology students raised the issue of sterility of the water samples. So much for "molecular" education at the expense of old-fashioned biology.

The work was conducted in a government Environmental Protection Agency lab, apparently seeking to demonstrate the presence of the specific reactive oxygen species in natural waters, thus perhaps with an unexpressed expectation that water pollutants would be oxidized naturally. By extension, it was suggested that the presence of humic substances in swamp water act as photosensitizer for generation of $^1O_2$, also that $^1O_2$ photochemical formation is widespread in aquatic environments. No correction of the paper has occurred. Zepp has abandoned $^1O_2$ studies but continues to work on water photochemistry.

Using these two examples of experimental design formulated by investigators to address a specific question, our class succeeded in more than one way. Beyond the shock to our students of seeing a prized classic paper with blemish and error but nonetheless with definitive proof of a major issue, one sees that the respected journal *Nature* publishes flawed, uncertain items as if they be so. Also, an imperceptible lesson was passed: the more one knows of a subject the easier it is to examine it for value, to understand it, to detect non-sequiturs, speculation, and error. The less one knows of a topic the easier it be to agree with matters seriously wrong, to overlook error, to accept the paper's design, results, and conclusions at face value, to be confused. Thus flourishes error!.

Where experimental design is flawed beyond recovery, it is necessary in some cases for retractions to be made. Such is more important when the flawed paper is to be used to effect public policy, medical treatments, and the like. The recent retraction of a report dealing with enhanced estrogen effects of pesticides on endocrine systems of bioengineered yeast cells is an example. The report of synergism became of importance to public policy regarding pesticide use. A flaw in experimental design was considered cause.[4]

5. Critical Assessments

The critical assessment of a published paper is absolutely essential to anyone seeking to know more about a problem, perhaps to use the methods, build on the results, or institute a whole new research program based on disclosures in one paper. From the imperceptible lesson of the first class it is inescapable that one must possess adequate knowledge of an issue before sound criticism is possible. Uninformed readers may always be led astray; critical evaluations so obviously depend on the state of the reader's knowledge and understanding of the science.

Three papers dealing with biochemical matters were

next assigned for criticism. Two articles were from the sacred five journals; one was outside this group. We often tried to get student remarks about the quality of the paper before discussion, and most students thought these papers were good ones, as the topics were close to their own interests and backgrounds. Only after revelation of the depth of error in each item did the students become more fully aware of the bother of having to read critically or risk being misled.

We regularly emphasized that uncritical acceptance of a method, a paper, conclusions, or concepts could set one back years. Reliance on a foolish concept, or on inappropriate or unreliable methods, could destroy the research plan of a naïve investigator who did not bother to confirm the suitability of literature guidance.

Given a paper for assessment, one must first determine what type of paper is it - an account of accomplished research, a full paper, a communication for rapid publication, a short paper with preliminary data perhaps published by photocopying the original typescript rather than by setting type, a letter to the editor, a news item, or a review. Each sort has its place; each must be evaluated for content but also for possible use..

Why should we read this paper; what use may we put to it. Read the title, the authorship, the locality where conducted; all are important ultimately to evaluation. A poor or misleading title, no matter whether it be from Harvard University Medical School or no, is still a poor start for a paper that may be important. From title on, one must be on guard for hints of inadequacy, error, or worse.

Papers generally begin with an introduction in which background and purpose are given. Full disclosure of methods follows in research papers. Abbreviated or bare mention of methods may occur in the shorter and preliminary papers, but in a full paper inadequate disclosure is a direct clue to reliability that must not be overlooked. Use of methods described in a journal article "in press" in slow-to-publish journals or in otherwise unavailable sources is a major warning sign limiting acceptance.

Anent results, there are many signs to be observed. What was done, what ought to have been done, proper statistics, graphs, regression coefficients, proper controls, assumptions, artifacts, and use of controls from prior literature of questionable value are among the many points to be examined. Discussion of results and proclamation of their meaning are vital items to be carefully evaluated, as it is here that major error may result. Sound data but results misinterpreted by naïve, ignorant, or arrogant authors must be discerned lest error be continued. It is here, of course, where the reader's own maturity is crucial; even senior investigators can be misled into serious error by published papers remote from their experience, outside their fields of expertise. We can all be misled, even duped!

Equally important to critical assessment of a paper is the citation of references used in the research. Experienced investigators ofttimes rely on the citation list as a prime means of evaluating a paper. Do the authors know the pertinent literature; do they cite true sources, or do they cite their own derivative works and ignore original work for improper advantage. Are unimportant items cited where important ones are not?

We limited this assignment to three papers that could be discussed in one class meeting. All three papers are flawed but in different ways. The first paper is from a membrane biology journal, the other two are from favorite "molecular" journals.

A. Klip, S. Grinstein, and G. Semenza, "Partial Purification of the Sugar Carrier of Intestinal Brush Border Membranes. The Enrichment of the Phlorizin-Binding Component by Selective Extractions", *Journal of Membrane Biology*, **51**, 47-73 (1979).

This paper has impressive credentials; it deals with an important matter of concern at the time, is from the laboratory of prominent European biochemist Giorgio Semenza, editor of *FEBS Letters* journal, and describes work done at the Eidgenössischen Technischen Hochschule, Zürich. These attributes notwithstanding, the paper is flawed in many ways. It was the selection of this paper by one of our better graduate students as basis for her student seminar that led us to create our course here described.

As already mentioned, a journal article title is an important item for critical evaluation. Here "the Sugar Carrier" in the title implies there be but one, thus is a matter for question. Moreover, we learn that the sugar carrier protein of the title is actually being sought using the nonsugar glycoside phlorizin and not the natural substrate D-glucose. The Summary of the paper contains the peculiar terminology "negative purification" and the undefined abbreviation NaDodSO$_4$, thus alerting us to be watchful for odd word uses.

The lengthy 27-page paper describes unsuccessful attempts to purify a brush border sugar transporter but fails to state explicitly that nothing of importance was accomplished. The paper is more a progress report in which the works of the two junior authors appear to have been joined to get something from their work. Given the many problems of the text, one suspects editorial privilege was extended to Professor Semenza. The many editorial flaws occurred despite the acknowledged aid of several who helped with preparation of the manuscript.

We are embarrassed to recount the many perceived problems of the paper, but it is a good medium for establishing once and for all the high cost of reading and evaluating flawed manuscripts, not necessarily texts in error. As suggested, peculiar word uses and jargon abound, among which are the examples: "spun down", "prerinsed" (rinsed says it all), "overshoot", "leakiness", "inwardly directed", "to pinch off", "fuzzy bands", and "fuzzy region".

There are the usual minor editorial problems, including several spit infinitives and misspelled word (sucrose for sucrase!). One datum is presented with standard deviation SD, another with standard error SE.

Words suggesting bias are used repeatedly, "notion" for idea or concept, "unfortunately" to pass judgement. Many qualified expressions creating uncertainty warn us to exercise care in accepting results and interpretations: "can be rationalized", "alternative explanations... are unlikely", "if it is assumed", "Assuming that", "suggests", "could indicate", "possible involvement", "can be explained".

Other expressions create uncertainty; protein was determined "In some cases". In Figure 1 we see results of a "typical experiment", always a remark that should be carefully noted. The graphed data have no error bars drawn, a matter instantly noted by our students. Results of two experiments with brush border vesicles are compared in Figure 2 in which "different batches of vesicles" were used, with no evidence that the batches had the same characteristics.

We are told three times, redundantly and to no worthwhile purpose, that the sugar transport protein is 0.4% of the total membrane protein, a value based on flawed estimates. Here, we recall Lewis Carroll: "Just the place for a snark! I have said it thrice. What I tell you three times is true".[5] I cited this last line in one of my papers that attempted to correct an erroneous assertion regarding $^{1}O_2$ reactions, repeated four times to no advantage in a paper dealing with the topic.[6] The editor asked me to delete the citation; at least he read my paper.

Critical assessment of this article must conclude that it be flawed, be a status report reflecting editorial privilege, and be of no further importance to the topic of the title. As regards experimental design, none of the approaches gave useful information or results that might provide basis for further investigation. Our students made several suggestions about how this problem might be examined, but their suggestions generally involved methods not available at the time.

The second journal article assigned for critical assessment is from one of the prestigious journals of the present dealing with molecular and cell biology matters.

T. Finkel and G. M. Cooper, "Detection of a Molecular Complex between *ras* Protein and Transferrin Receptor", *Cell*, **36**, 1115-1121 (1984).

This paper too has impressive credentials, coming from Harvard University, Harvard University Medical School, and the Dana-Farber Cancer Institute. The paper addresses a topic of major concern at the time, the role of Ras proteins in cellular activities, perhaps in the regulation of cell proliferation. *In vitro* immunoprecipitation of transferrin receptor protein with monoclonal antibodies raised against Ras proteins erroneously suggested formation of a Ras protein-transferrin receptor protein "molecular complex", thus that Ras proteins regulate cell growth by interacting with the cell surface transferrin receptor.

We are alerted at the top of the first page that less than full care is possible in this "advertisement". The April 1984 journal issue is given the volume number **136**, where Volume **36** is correct. With such editorial assistance it is not easily determined whose error it be in the text of a published article.

As with other titles of papers, the present title raises an issue neither addressed nor resolved. Here the term "molecular complex" is in the singular, so only one such complex is suggested, no matter how many Ras proteins there might be. What may seem a minor matter of composition is actually of deeper philosophical concern; how many Ras proteins are there, and how many complexes.

Also, right there on the first page we see the frequently encountered evidence of ignorance of the meaning of words; the molecular weight of proteins is expressed in "daltons" (not the capitalized Daltons, a unit of mass, not weight; see Chapter 4). With this and the title imperfections one is now on guard that serious error may exist, that the authors do not realize or care about error. Pressures to publish are simply too great to be concerned with such matters.

Once on guard, one sees many minor items that alert the attentive reader to possible problems, some serious. Among such bothersome items is a full use of laboratory jargon; e.g. "were passaged", "flowthrough fraction", "to preclear", and "precleared". Some chemical symbols (NaCl) are used, but there are also spelled names (ammonium sulfate). The popular excuse of a protein "conformational change" is advanced as explanation for inadequately examined events not really understood for lack of evidence.

These quibbles aside, it is the experimental design itself that is seriously flawed. A goat or rabbit anti-rat immunoglobulin G was used, there being unrecognized transferrin present in the commercially available material, transferrin that then bound the transferrin receptor, to produce the precipitate encountered. Moreover, improper controls were used, as normal rat serum was used in control work in place of the monoclonal antibodies. The serum contained sufficient transferrin to prevent receptor binding. Nonspecific serum should have been used in any event.

Two common errors occur here. Inadequate or improper controls are too often encountered, thus making experimental results uncertain or in abject error. Secondly, there is undue reliance on commercially available preparations without careful analysis for their composition, expected activity, required effectiveness, and absence of other impurities or agents potentially affecting the outcome of an experiment. Buy it, use it, interpret odd results as best you can, include "conformational change" and heroic claims of new *in vivo* meaning.

The perceived importance of the claims made is suggested by the rapidity with which the paper was published, in April following receipt February 17 and in revised form February 27. This very fast action suggests special consideration; we note the journal *Cell* editorial office and Harvard University are close by one another. We also note that the two authors were no longer affiliated with Harvard University in October following the disclosure of their error in the third paper under study next.

The third paper of this lesson (handed out in class) is also from among the favored five journals of molecular biology. The paper in *Nature* reexamines the claims of Finkel and Cooper published earlier the same year in *Cell*, and exposes their shortcomings for all to see - an object lesson in how error be exposed to the embarrassment of the perpetrators.

J. Harford, "An artefact explains the apparent association of the transferrin receptor with a *ras* gene product",
Nature, **311**, 673-675 (1984).

The occurrence of error through poor experimental design leads not only to misleading published articles but also to the necessity for corrections. Unfortunately, others may have to recognize the error, conduct experimental work to test matters, and then disclose the error, at times to the dismay of the scientists making the error.

Such is the present case. Harford carefully describes the error of Finkel and Cooper and provides sound experimental work to show the error caused by inadequate control work. There are no oddities in the paper that attract adverse criticism, the only odd matter being citation of the Finkel and Cooper paper as volume **136** of the journal *Cell*, perpetuating that error.

These papers presented experimental design in three different ways: unsuccessful approach leading nowhere, bad control work, and a correction of error with sound experimental design. Our students generally were confused about what to say about the membrane sugar carrier paper and were unable to spot the glaring error of the Ras protein-transferrin receptor paper that was more to their liking. Only with the third paper was there general acceptance of good experimental design.

6. Reviews

The use of suitable review articles for ready orientation to the background of a topic or concept, to suitable methods of investigation, and to current status is of great importance. All that is needed is selection of the appropriate review, but such selection is not always apparent from author, title of paper, or journal source. There are excellent reviews, good reviews, flawed ones, and poor ones, in highly regarded journals but also in other sources.

The labor of writing a sound professional review article, even of a limited topic, is not apparent to those who have not tackled such a job, and is probably beyond student grasp. As a prelude to the writing assignment later in our course and for naïve entry upon a new topic, selection of a proper review article is essential. There is merely the issue of how to find reviews, evaluate them, and select those that are suitable.

We presented three reviews for student criticism, one essentially useless but in a publication suggesting value, one trivial but in an important medical journal, and one a detailed review in a highly regarded biochemistry journal. Our selections were designed to disclose problems in some reviews but also, insidiously, to reveal that excellent reviews may not be readily identified. We also had to take into account the length of the review lest too great a length compromise heuristic value.

The first review paper dealt with the potential carcinogenic activity of hydrazine derivatives and included historical background material and personal experiences of the author.

J. Baló, "Role of Hydrazine in Carcinogenesis", *Advances in Cancer Research*, **40**, 151-163 (1979).

The review published in a hardbound Advances volume emphasized the work of Baló and his students. Despite the 1979 date of the article, the latest items cited were from 1976, with most well before that, whereas many later papers and items were available for discussion. Moreover, the highly selective nature of items discussed and cited, together with emphasis on the author's work and excessive details about trivial aspects limited the review at the time of publication to historical perspective and, as in our use of it, to instruction how not to select or write review articles.

Again, the title reveals problems to follow, in this case

in word usage. The title should have been hydrazine derivatives, not just hydrazine, as emphasis is not on simple hydrazines but on hydrazides and nitrosamines as well. As noted by Mark Twain, there is a "difference between the lightning bug and the lightning". Hydrazine is called a diamide. Hydrazine and its hydrochloride salt, but not the dihydrochloride, are mentioned. There are trivial matters; "gm" is used for gram "g".

More serious matters include assertion that isonicotinic acid hydrazide (isoniazid, INH), a drug of choice for treating tuberculosis, is carcinogenic! Total misconceptions are expressed: "which part of the INH compound was carcinogenic". Peculiar metabolism pathways for dialkyl hydrazines are proposed; organ-specific hydroxylases metabolize the compounds, and cancers due to hydrazine occur in tissues where it is metabolized. Outstanding is "the conclusion is inescapable that all hydrazines are carcinogenic with the possible exception of methylhydrazine"! Such remarks would not appear in a serious review. One suggests editorial privilege be involved in this case, as the review has no usefulness save for criticism.

The short second assigned review emphasized a particular *in vitro* bioassay system advanced as test of environmental carcinogens, a matter of controversy at the time.

C. Heidelberger, "Chemical Carcinogenesis", *Cancer*, **40**, 430-433 (1977).

Once more, the title must cause pause for reflection, as in no way does the article cover such a broad topic. We see a farago of homilies, history, amusements, and promotion and enthusiasm for a simple bioassay system. Percival Pott, 1775, is invoked as the initiator of cancer research. Rapid career through generalized concepts of chemical carcinogenesis bring us to rejection of the Ames *Salmonella typhimurium* bacterial test for mutagenicity, as "*Salmonella* do not get cancer", in favor of an *in vitro* eukaryote test system devised by the author. The review closes with clear warning that bias has been expressed: "I hope that I have convinced you of three things".

Our students were confused by this brief review, as Heidelberger cited a more thorough review of his as background. Of course, the review is a review because we said it was. In reality it was an after-dinner speech presented at an American Cancer Society meeting and is hardly a review at all. The nature of this address was not recognized by the students despite a footnote making the matter certain. Why read footnotes; they only distract!

The third assigned review was a detailed treatment of a narrow topic but one involving regulation of an enzyme of importance in biomedical studies.

P. J. Kennely and V. W. Rodwell, "Regulation of 3-hydroxy-3-methylglutaryl coenzyme A reductase by reversible phosphorylation-dephosphorylation", *Journal of Lipid Research*, **26**, 903-914 (1985).

Students regularly found this high quality, fully professional treatment of the topic to be too detailed, they not realizing from prior indoctrination course work that great detail is what it takes to characterize an enzyme. Students used to easy descriptive material have no comprehension of enzymology and lose interest in such detailed biochemistry in any event. We got the same responses in prior classes when an earlier review of the same enzyme but in a less prestigious journal[7] was assigned. The prestige of the journal source was of no consequence; the review was just too detailed!

Here the title offers no hint of problems to come but succinctly tells what is involved. An outline of topics begins the review; characterization of the enzyme, its role in the sterol and isoprenoids biosynthesis, and its regulation by reversible phosphorylation follow. Other than too many split infinitives and the expression *raison d'etre* not italicized nor recognized by our students, there are no glaring composition problems.

The review ends with discussion of points bringing the

reader immediately to the status of the subject at the time of publication. In the face of convincing arguments for the reversible phosphorylation mechanism, there was then opposition to such regulation being physiological. The unnamed proponents of this opposing notion did not see a need for short-term regulation, given their own formulation of genetic regulation mechanisms for which Nobel Prizes were awarded them. The authors conclude that they do not know why the enzyme is so regulated but that such regulation is the case.

The 1985 review includes 100 references, eleven of which are from 1984, four from 1985, and some "in press". Such up-to-date coverage establishes the authors as experts on their review topic and secures the review as a valuable contribution of the time. Subsequent events have not discounted the reversible phosphorylation regulatory mechanism.

7. Special Topic. In Vitro Transcription

In our second set of lessons we careered in greater detail among specialist topics of the interests and expertise of the instructors: *in vitro* transcription in the molecular biology realm, oxygen biochemistry in a biochemistry realm. The assigned papers were up to thirty years old, and students often remarked that they would rather have more recent papers on more timely subjects. However, the lessons to be learned were too good to avoid use of these items. Furthermore, in this regard we admonished students that we did not seek to add unnecessary details to the assignment nor to make them instant experts on the topics. Rather, these special articles were selected for the lesson they provided in making points very solidly, also to capitalize on the instructors' special interests and expertise. Having access to the most advanced background on these special topics was necessary to delve into the deeper problems of concepts and methods, even though each special topic has passed from current interest.

In vitro transcription studied with operationally defined nuclear chromatin composed of DNA, histones, and perhaps other proteins was of great interest at one time. The topic is well within the experience of molecular biology students, and the material was mastered easily. Yet despite a grasp of modern methods and concepts these first papers still posed difficulties in interpretation. The first assigned paper was:

> J. Paul and R. S. Gilmour, "Organ-specific Restriction of Transcription in Mammalian Chromatin", *Journal of Molecular Biology*, **34**, 305-314 (1968).

The paper from the respected Beatson Institute for Cancer Research attempts to show that histones binding chromatin DNA do so non-specifically. The authors recognize the primitive nature of their methods and write cautiously. Chromatin was defined operationally but without proper characterization. Terms describing it include "dehistoned chromatin", "clean preparation", "reproducible chromatin preparations", and "essentially orthodox", all leaving one unsatisfied as to just what was prepared. Moreover, many qualifying terms are used in the discussion: "most likely interpretation", "second assumption", "generally assumed", "circumstantial", "latter speculation", "exercise caution", "suggests", "indicates", "not too unlikely", "suspected", and "postulated". What are we to conclude?

The second paper, two authors of which now members of the National Academy of Sciences, continues matters:

> R. Axel, H. Cedar, and G. Felsenfeld, "Synthesis of Globin Ribonucleic Acid from Duck-Reticulocyte Chromatin In Vitro", *Proceedings of the National Academy of Sciences USA*, **70**, 2029-2032 (1973).

The article attempts to show that chromatin proteins restrict transcription in a very specific manner; the authors pose transcriptional control factors in eukaryotes. Despite the authors prominence, laboratory of origin, and journal involved the study has serious errors in experimental design that effectively destroy value. Nonetheless, the paper may be discussed in great detail in class to advantage.

Among poor experimental practices was the addition of RNA preparations to chromatin preparations without regard for compartmentalization bound to occur in the multiple-phase system. Bacterial RNA polymerase was used to form eukaryote RNA. Controls were conducted differently from the experimental work. Different levels of newly synthesized RNA protect endogenous RNA from RNAse actions. Despite the end assertion of the authors "our results appear to provide convincing evidence of transcriptional control factors" the words "appear" and "convincing" tend to cancel one another in meaning. The paper is now judged to be worthless.

The topic was of continuing interest at the time, and other approaches in which mercurated substrates for chromatin were utilized:

G. F. Crouse, E. J. B. Fodor, and P. Doty, "In vitro transcription of chromatin in the presence of a mercurated nucleotide", *Proceedings of the National Academy of Sciences USA*, **73**, 1564-1567 (1976).

This paper from Harvard University examines *in vitro* transcription using uridine triphosphate chemically modified by covalent bonding of mercury. Again bacterial RNA polymerase was used to prepare transcripts, the organomercury transcripts then being separated from nonlabeled transcripts. Controls were conducted differently from the experimental work. Although the authors recognized aggregation of RNA as a problem, their management of the problem is inadequate. The authors qualify their work with "We have therefore relied on" work of others and use peculiar terms "infinite time" that add misunderstanding.

Following three papers with serious errors in experimental design it was necessary for more thoughtful studies to be crafted to correct perceived error. My biology colleague had entered into his research career at this point with:

D. A. Konkel and V. M. Ingram, "RNA aggregation during sulfhydryl-agarose chromatography of mercurated RNA", *Nucleic Acid Research*, **4**, 1979-

1988 (1977).

Because of the then extensive importance of *in vitro* transcription rapid publication in a journal using photographic reproduction of a typed text was deemed important, thereby to alert investigators of the serious problem of RNA aggregation.

The uncontrolled aggregation of mercurated RNA in prior studies was revealed. The paper is flawed by nonce words, jargon, and undefined acronyms, and is a good example of what can be wrong in rapid communications printed without opportunity for proper editorial review. In one place DNA is written where RNA is correct. There are three different ways of expressing the state of Massachusetts, just a minor matter but one clearly revealing the pitfalls of rapid publication.

Konkel and Ingram were not the only ones who alerted readers to the RNA aggregation limitation, witness:

M. Zasloff and G. Felsenfeld, "Use of Mercury-Substituted Ribonucleoside Triphosphates Can Lead to Artifacts in the Analysis of In Vitro Chromatin Transcripts", *Biochemical and Biophysical Research Communications,* **75**, 598-603 (1977).

This paper from the NIH, also a rapid communication, outlines another error in the use of organomercury nucleotides. In keeping with nonce fashion there is the assertive sentence title that tells us what to believe. However, the authors do recognize the problems of "chromatin" and "purified" RNA product. The work contains some experimental problems (bacterial RNA polymerase was used), but conclusions are correct. A full journal article eventually appeared.[8]

Finally, we assigned a paper from the Beatson Institute for Cancer Research that continued studies of *in vitro* transcription plagued with unrecognized aggregation problems.

R. S. Gilmour, M. Allan, and J. Paul, "Accurate initiation of human ε-globin RNA synthesis by

Escherichia coli RNA polymerase in isolated nuclei of K562 erythroleukemia cells", *Proceedings of the National Academy of Sciences USA*, **81**, 4051-4054 (1984).

National Academy of Sciences member Gary Felsenfeld submitted the paper on behalf of John Paul, Fellow of the Royal Society.

The topic is well within the experience of students but again involves operationally defined chromatin, with no way to tell what is in each preparation. Adequate controls were not conducted, and the 1977 paper of Konkel and Ingram alerting to aggregation problems was not cited despite its ready availability. The by-now extensive use of molecular biology acronyms, jargon, and gibberish included upstream, K562 cells, *Mbo* II DNA, Hg-UPT, pATH εG, *Eco*RI, pAT153, and "TATA" box.

8. Specific Topic. Oxygen Biochemistry

We touched on two topics absolutely crucial to living systems, including the favored cell cultures of the molecular and cell biologists. However, neither oxygen discussed here nor water discussed under Polywater is a controlled entity whether *in vivo* or *in vitro* studies be contemplated. Accordingly, one merely ignores the problem of any influences that water activity and reactive oxygen species might have on the system observed.

A topic of great interest 25 years ago was the question whether a particular excited diatomic molecule of oxygen, (singlet oxygen, $^1O_2$), form in living systems and oxidize tissue components via controlled enzyme reactions or by uncontrolled reactions potentially causing harm to the system, including lipid peroxidations.

Again, despite the age of the topic, the lessons to be gleaned are too sound to have topic age deny access. In reality, $^1O_2$ has been confirmed by more modern methods as a minor component in some *in vitro* biochemical reactions, but the significance of the matter remains obscure and uncertain for *in vivo* studies and in studies of the natural environment.

The reading assignments were treated in two classes, with three assigned papers for each class. Each set of papers was usually more than enough to fill class times with lively discussions. The first set was to orient the students to some of the difficulties in approaches to study of oxygen biochemistry. In the second class, we attempted to resolve some of the problems arising from the first set of papers.

The first assigned paper offered an alternative notion about enzyme activity in sequestering $^1O_2$ *in vitro*:

> W. Paschen and U. Weser, "Singlet oxygen decontaminating activity of erythrocuprein (superoxide dismutase)", *Biochimica et Biophysica Acta*, **327**, 217-222 (1973).

In this paper (one of several) Paschen and Weser attempt to show that the blood protein erythrocuprein intercepted and destroyed $^1O_2$ *in vitro*, thus providing a previously unrecognized physiological function for erythrocuprein. The authors went so far as to create a name "singlet oxygen decontaminase" and acronym SOD for erythrocuprein to replace the then standard "superoxide dismutase" (SOD) terminology.

This peculiar correspondence of the acronym SOD with the two different meanings is an example of a well-known method of advancing one's thesis "Name It And It's Yours" discussed in Chapter 5. By assigning a name to a phenomenon one establishes its true existence and ensures proponents their moment in the sun. In this case, by sequestering the acronym SOD for $^1O_2$ decontamination there developed a direct affront to the proven superoxide (O_2^-) disproportionation evidence.

Erythrocuprein was known for years as a copper-containing plasma protein but with unknown function. However, this now older material was no longer presented in the condensed biochemistry course, so our students knew nothing of erythrocuprein, the true physiological function of which is the disproportionation of O_2^-, the one-electron reduction product of dioxygen O_2. This identity was well accepted at the time, but there had been speculation that $^1O_2$ formed in biological systems may have arisen from O_2^-. Thus, there was choice which formal process be correct.

$$2O_2^- \rightarrow O_2 + O_2^{2-} \text{ (dismutation)}$$
$$^1O_2 \rightarrow O_2 + hv \text{ (decontamination)}$$

Perhaps better questions be posed: was either process valid? Could one be valid and not the other? Could both be valid?

Our students had been exposed to the error of using nonspecific interceptors for $^1O_2$ in our first class dealing with swamp water, but the Paschen and Weser paper took them beyond this experience. However, by this time our students were willing to grasp possible prior errors accepting SOD activity against O_2^- in favor of the $^1O_2$ decontamination concept.

This developed into another little trick by us to force student attention to details beyond their experience. Students still did not check references cited for background or explanation of points not understood. Students also failed to learn the meaning of words they encountered but did not know, in this case "entatic" used three times! Terms such as "gatoms" went unrecognized by our students.

The key to the present error lay in selection of analysis procedures, coupled with misunderstanding of the relevant basic chemistry. Among the several errors of the paper was the reliance on the reaction of potassium peroxychromate K_3CrO_8 with water to yield $^1O_2$ exclusively, with the effect of added SOD being to inhibit emission of visible light from the coupled oxidation of luminol, thus an assay for $^1O_2$ in a mixed assay system of some complexity. The authors stipulated that $^1O_2$ was formed exclusively from CrO_8^{3-} decomposition, and that they had "strong evidence that no superoxide at all was generated", so they were aware of the possible error but erred in not being more certain from experiments in their own laboratory Both $^1O_2$ and O_2^- are formed, but only $\leq 6\%$ is $^1O_2$.[9,10]

Additionally, the superoxide dismutase bioassay was conducted with a cytochrome c reductase but using potassium superoxide (KO$_2$) in dimethylsulfoxide as O_2^- source. The alkalinity of KO$_2$ was not considered, neither was the possible influence of both $^1O_2$ and O_2^- in the system. It so happens that this assay system is flawed, as $^1O_2$ oxidizes Fe(II) cytochrome but O_2^- reduces Fe(III) cytochrome, thus yielding a meaningless analysis system that may merely reflect the balance between $^1O_2$ and O_2^- fluxes. Also contributing to uncertainty is the report that O_2^- quenches $^1O_2$,[11] so in systems where both may exist additional control must be exerted. Thus the claim that SOD intercept and relax $^1O_2$ but not disproportionate O_2^- and the assertion of "unequivocal evidence" that O_2^- not be involved in $^1O_2$ production be pure error, exactly of the sort we sought to dispel in our students future experimental work.

Further search for $^1O_2$ within *in vitro* biological systems took other approaches:

M. Nakano, T. Noguchi, K. Sugioka, H. Fukuyama, M. Sato, Y. Shimizu, Y. Tsuji, and H. Inaba, "Spectroscopic Evidence for the Generation of Singlet Oxygen in the Reduced Nicotinamide-Adenine Dinucleotide Phosphate-dependent Microsomal Lipid Peroxidation System", *Journal of Biological Chemistry*, **250**, 2404-2406 (1975).

In this Communication we see an attempt to demonstrate formation of $^1O_2$ in the specific *in vitro* NADPH-dependent rat liver microsomal lipid peroxidation system so popular at the time. In the Summary we see the light emitting species was "estimated to be singlet oxygen", thus a qualified conclusion.

The method used had technical shortcomings; emitted

light was recorded using filters in discrete wavelength ranges. Data were reported in Ångstroms (Å), an unknown measure for our students. Visible bands at 5200 Å, 5900 Å, and 6350 Å were recorded and compared with emission spectra of $^1O_2$ formed by reaction of H_2O_2 and NaOCl, an established system generating $^1O_2$ when properly conducted. Because of instrument limitation an intense band at 7032 Å could not be observed in either system. This bothered our students, as they had no experience with emission spectra and could not agree that the spectra presented were good matches with one another, given the method used.

A more serious flaw in the paper was that attention was given to a time after initial events had occurred. Given that there was not enough time for emission spectra to be recorded rapidly in the early phases of the experiment, the authors studied light emission after 10 minutes (with maximum light at about 58 minutes) but did not do so for the first 10 minutes following addition of NADPH that initiated the peroxidation process. However, there was significant light emission by total counts that increased to a maximum at about seven minutes but subsided by 10 minutes, after which time emission bands were detected with the spectrometer filter system. Now, if $^1O_2$ be generated following initiation of lipid peroxidation, thereby to be a significant agent in peroxidations, one might expect early or direct formation in the first phase noted, not in the longer second phase.

Catalase and SOD were without effect; 2,5-dimethylfuran as an efficient but nonspecific interceptor of $^1O_2$ inhibited light emission. The authors conclude that it is unlikely that either O_2^- or hydroxyl radical HO˙ be the source of $^1O_2$ produced in this system. This conclusion is probably correct, but whether $^1O_2$ was formed at all remains uncertain.

A third assigned paper presented a much greater detailed study of an *in vitro* enzyme system:

H. Rosen and S. J. Klebanoff, "Formation of Singlet Oxygen by the Myeloperoxidase-mediated

Antimicrobial System", *Journal of Biological Chemistry*, **252**, 4803-4810 (1977).

In this lengthy and detailed paper the authors attempt to demonstrate the enzymatic formation of $^1O_2$ in incubations of myeloperoxidase with H_2O_2 and Cl^-. The study suffers from the same fatal error of Zepp *et al.* using a furan as $^1O_2$ interceptor previously discussed, as the diphenylfuran interceptor used is not specific for $^1O_2$. The authors remark that a "relatively unique product may be formed". Whatever oxidant be generated in their myeloperoxidase system remains uncertain.

The authors were aware that other interpretations of their results may be had, that it be HOCl formed and not $^1O_2$, but they were unaware of the failure of furan interceptors to sequester $^1O_2$ specifically. As in the Zepp *et al.* paper, many data were adduced in support and for verisimilitude. However, the paper oozes with repetitive assertions and wishful thinking regarding interpretation of results. Remarks such as "all the evidence taken together suggests strongly that $^1O_2$ is formed" fail to convince, as compounding several uncertainties cannot accord a certainty. Our students grasped this problem of assertions in error and began to disbelieve. Preconceived results, bias, repetitive assertions, and wishful thinking, all items of uncertainty, were summed to a conclusion of certainty.

In the second class papers were assigned that reexamined the problems just identified:

A. P. Schaap, A. I. Thayer, G. R. Faler, K. Goda, and T. Kimura, "Singlet Molecular Oxygen and Superoxide Dismutase", *Journal of the American Chemical Society*, **96**, 4025-4026 (1974).

As with the *in vitro* transcription papers needing corrections for errors in experimental design, so also the previously discussed claim of Paschen and Weser that erythrocuprein be a $^1O_2$ decontaminase and not a superoxide dismutase required correction accorded by the present article. Here α-lipoic acid was used as interceptor of $^1O_2$ in two systems generating $^1O_2$ reliably, with and without added SOD enzyme. The extent of substrate oxidation was assessed by changes in absorption spectra. In a photochemical system generating $^1O_2$ SOD had no effect; controls established that a recognized specific interceptor of $^1O_2$ did reduce the extent of oxidation and that the system did not destroy SOD activity.

In a second system an adamantane ozonide derivative generating $^1O_2$ upon warming likewise was unaffected by SOD. Here the ozonide at $-10°C$ and the aqueous test solution were mixed and allowed to warm to room temperature in 15 minutes, but no control demonstrating that SOD retained activity under this condition was mentioned. In neither case was the $^1O_2$ oxidation of α-lipoic acid diminished by SOD, thus settling this matter definitively.

The question of $^1O_2$ formation by myeloperoxidase was also addressed:

J. E. Harrison, B. D. Watson, and J. Schultz, "Myeloperoxidase and Singlet Oxygen: A Reappraisal", *FEBS Letters*, **92**, 327-332 (1978).

Perceived published error begets correction papers. Here the prior conclusion of $^1O_2$ formation in the myeloperoxidase system is reexamined, and additional experimental evidence is adduced in favor of HOCl formation, not of $^1O_2$ formation. Other than minor editorial blemishes, split infinitives, "This data", and the like, the paper clearly adjusts the matter.

Others have since confirmed that $^1O_2$ formation is negligible in the myeloperoxidase system at pH 4-5, but the identity of the oxidizing species implicated is dependent upon pH of the system, Cl_2 being the agent at low pH, Cl_2 and HOCl at pH 2-3, HOCl and ClO$^-$ at neutrality, and ClO$^-$ at high pH.[12]

Our students had some difficulty understanding the technical nature of the experiments described, as again spectral data were presented. Also, the relevant chemistry was beyond their interests. However, grasp of the lesson was had, that corrections of honest error may be made by others.

Finally, a seminal paper reliably demonstrating $^1O_2$ in an enzyme system was assigned:

J. R. Kanofsky, "Singlet Oxygen Production by Lactoperoxidase", *Journal of Biological Chemistry*, **258**, 5991-6003 (1983).

It was our fashion to assign papers without serious problems for evaluation, thereby to force students to work a bit. We selected this paper dealing with oxygen biochemistry by a physician who sees patients and conducts research at a Veterans Administration Hospital. The assignment once more took our students outside of their experience, to their consternation.

Prior methods for detection of 1.27 μm chemiluminescence light emitted from biological systems releasing $^1O_2$ were unsuitable for reliable use, witness the Nakano *et al.* article just described. In the present seminal communication Kanofsky described an advanced photocell detector for 1.27 μm light and in so doing created means of convincing demonstration of $^1O_2$ in biological systems, means now *de rigueur* for such work.

Our students all doubted this article's value, as none was acquainted with near infrared emission spectroscopy, and by now the students were looking for error rather than evaluating papers on merits. Only minor editorial flaws can be identified: use of the term "half-life" where "lifetime" is correct, and misuse of "since" where "whereas", "as", or "in that" be proper use. However, an insignificant technical flaw is noted, just to establish that nothing is perfect. Kanofsky cited $^1O_2$ lifetimes of 2 μsec in ^1H_2O, 20 μsec in ^2H_2O where better values were known at the time (3.3-4.4 μsec in ^1H_2O, 55-58 μsec in ^2H_2O), suggesting unawareness of recent literature developments.[13]

Our students did not like to accept 1.27 μm light

emission as reliable evidence. Their discomfort with spectroscopy coupled with our harsh treatment of error in assigned work made them wary of new methods. How can we be sure the 1.27 μm light is specific for $^1O_2$? Just because you say? No, because the region is vacant of other known spectral emissions except for that of $^1O_2$.

These assignments taught imperceptibly several important lessons that we did not need to emphasize. Early exploratory work in an emerging field is fraught with the potential for error, all in ignorance. Again, if you do not know anything about it, how can you judge. Even with a bit of background, but not expert status, should you venture a strong opinion. Perhaps a modest stance about any new science item is better.

On occasion two charges were levied against such in depth treatment of oxygen biochemistry. The topic is too far from "molecular" matters, is too much like chemistry, and the papers are dated. Whether awareness of basic chemistry is of value to "molecular" investigations or no, the topic of reactive oxygen species in biology is now burgeoning, and vital links of reactive oxygen species to cellular function are being discovered. Just three items should serve to show relevance: control of gene expression and of cellular function by oxygen species is suspected,[14,15] HIV-1 promotor activation may be mediated by oxidative stress,[16] and current interest in apoptosis, programed cell death, now associates reactive oxygen species as possible causation.

Anent charge that our assigned papers be dated, the basic problem of uncritical use of fashionable and convenient methods to explore lipid peroxidation and oxygen damage in biological systems continues. A marked increase in such studies is now the case, but critical selection of reliable methods is not always had. Indeed, as in so many studies where fashionable but questionably reliable methods are used to hasten results, error is introduced.

In some cases it is also important to await resolution of

difficulties before advancing a paper, method, or results to the classroom for criticism. Direct criticism and immediate resolution of many matters is possible, but some are more refractory and require additional investigation not always leading to resolution.

The issue of suitable methods for a given study is ever important, and many poor choices have been made. As problem cases, recent instances involving the burgeoning study of biological oxidation phenomena are noted here. The popular use of lucigenin (L^{2+}) chemiluminescence in the detection and measurement of O_2^- in biological systems *in vitro* has been questioned as potentially unreliable. The one-electron (e^-) reduction product of lucigenin (radical ion $L^{+\cdot}$) might then react by redox cycling with O_2 to generate O_2^- in systems where none is present absent lucigenin. Morever, competition between lucigenin and O_2 for reaction with e^-

$$L^{2+} + e^- \rightarrow L^{+\cdot}$$
$$L^{+\cdot} + O_2 \rightarrow O_2^- + L^{2+}$$

might also occur, making measured results highly system dependent. Clearly, the most careful verification of the suitability and control of such methods is necessary.[17] A second continuing convenient means of preparing malondialdehyde (O=CH-CH$_2$-CH=O), not now generally available commercially, as a reference standard in measurements of its formation in lipid peroxidations and of its biological properties, is in the hydrolysis of 1,1,3,3-tetraethoxypropane ((C$_2$H$_5$O)$_2$-CH-CH$_2$-CH-(OC$_2$H$_5$)$_2$). However, the presence of biologically active ß-ethoxyacrolein (C$_2$H$_5$O-CH=CH-CH=O) is now recognized in such unpurified malondialdehyde preparations.[18]

Such common practice of using unpurified hydrolysis mixtures is typical of many biological protocols, where chemicals from whatever source are taken without consideration of identity or purity. Not only may chemical assays be in error, but bioassays may be greatly affected.

Depending on the sophistication of the class a selection

of one of these problems may be in order. However, topics dealing with real chemistry but not with descriptive matter generally are unsuccessful in making sound points.

It cannot be overemphasized that reliance on labels with regard to identity, purity, and amount is unsafe where serious work is to be conducted. My own experience includes two examples of problems derived from unwise trust in such matters. In a planned synthesis of the antibiotic puromycin I unwisely trusted the label of a commercially available tyrosine derivative, only to fail to get the sought product. Trouble-shooting established the problem: the starting material was something else, not what the label stated. Although my generation of chemists was trained to confirm identity and purity of reagents, even solvents, to be used, my lapse wasted perhaps a week of my time.

Secondly, in my absence from my laboratory, a bioassay of an as-delivered sample of an oxidized steroid was conducted, the test showing unexpected positive results. In attempted confirmation of the test the sample was repurified and other samples acquired and tested, all resulting in negative responses after much wasted effort.

Others of our experienced faculty had similar difficulties in extensive studies of biologically active materials in which strange, disparate responses not reconciled with other experience were obtained. In one case a marine toxin preparation consisting of several bioactive agents was unwisely studied; in another a commercially available thiazole derivative that was clearly decomposed (brown color!) had been used. Only at attempted publication were these errors of analyte identity and purity discovered from remarks of referees. Other such difficulties are discussed elsewhere in this Chapter.

Reliance on commercially available buffer solutions is frequent. These may be used even without checking the pH of the system. Yet another case involved a technician seeking from me an unopened one-pound bottle of trichloroacetic acid. The bottle must be unopened. Why? So a 10% solution could be prepared without weighing the material!

For those involved in mammalian cell culture the identity of cells obtained by gift or transfer from some other laboratory may not be confirmed by the same rigorous testing that microbiologists have had to use. Also, studies with mammalian cell cultures containing microbial growth as well is not always considered or detected.

9. Polywater

Following criticism of the two specialist topics *in vitro* transcription and oxygen biochemistry respite from specialty details seemed in order. However, our use of two additional topics of government funding frenzy and of the publication policies of journals was not without risk, as these items took our students well outside their experiences and interests. The saving grace was that two crucial lessons were to be learned, lessons neither obvious nor anticipated by our students.

For the first master lesson we chose the Polywater episode of 1966-1973 in which band-wagon after band-wagon rolled, at great government expense. We mention once more the importance of water and oxygen essential for life, both factors regularly uncontrolled and ignored in cell culture *in vitro* studies of so much current importance.

In 1962 Russians reported a new form of water, ordinary H_2O changed by peculiar surface effects in capillary tubes to a substance called water II or anomalous water, later "Polywater". Once again, giving the substance a name "Polywater" made it so. The news made domestic newspaper headlines[19] and shook the world 1966-1973. The episode erupted onto the scene in the mid-1960s, during the late "Cold War" period between the USA and the Soviet Union, apparently when scientists were asleep, the federal government alarmed, and the world unaware of imminent danger. Fear of Polywater danger prompted "chicken little" warnings in prestigious journals. Recall this was not long after Sputnik and the Kennedy Missile Gap charade of the 1960 presidential election, and now the Russkies were ahead of us on this too.

Not only would Polywater be in the same class as Watson and Crick and noble gas reactivity, but military interests were also paramount. If submarines were coated with Polywater they could run underwater with less friction and sound, more efficiently, and it would be Russian submarines if they beat us to it. Big federal funds were thrown into crash research.

Frenzy is an apt word describing Polywater work. Not like the genuine frenzy that occurred in December 1941 when a chemist discovered his laboratory in great disarray. It seems Army Chemical Corps chemists had used the laboratory to determine how to make sulfuric acid from chemical artillery shells for failing batteries on Bataan.

The new form of water was declared a water polymer of peculiar stability. Appeal was immense; all manner of conceptual or theoretical support was adduced. Although Polywater was never described adequately as a chemical, its structure as a water polymer was explained by deluxe theoreticians obviously seeking credit, and big government handouts. The bandwagon rolled. In fact, Polywater was no such polymer but a hydrosol with all the properties of same.

We took a chance on assigning our students research papers dealing with Polywater, as there are so many, so far outside the "molecular" experience, and so confusing even to the experts at first blush. On balance our students rose to the occasion and were able to grasp, not the technical features of the experience for want of background in physical chemistry, the overall lesson, that government can and does go to extremes where driven by politics, domestic and foreign.

For this class we usually assigned five papers. This was a heavy load in that our students were hardly prepared to criticize physical chemistry and spectroscopy papers. Nonetheless, we plowed directly on just to see the effects on biology students reading of a serious issue into which hundreds of scientists were drawn, along with unrevealed amounts of federal research money.

Work published in Russian from 1962 finally reached the western world in 1966. The seminal paper alerting the world of the Russian work is:

B. V. Derjaguin, "Effect of Lyophile Surfaces on the Properties of Boundary Liquid Films", *Discussions of the Faraday Society*, **42**, 109-119 (1966).

This paper is an English language review of Deryagin's earlier work on water preparations condensed in capillary tubes. As such the review made the western world aware of the discovery and touched off major efforts to study the material in the USA. The material presented by Deryagin involved several physical chemistry methods and presented results suggesting a possible new form of thermodynamically metastable water variously termed specific water, anomalous water, or orthowater to distinguish it from metawater, usual water, or normal water.

The paper introduced words and concepts not well received by our students, but by this time they had found out what "Lyophile" in the title meant. Words in the text such as anisotropic and piezoquartz went not understood, as did also the concepts of shear modulus, vapor pressure, and viscosity. When challenged to look at Figure 10 showing a freezing point curve for water and for anomalous water, the students could agree that a depression in freezing point was indicated for the specific water, this typical of aqueous solutions of salts. So much for early evaluation of Deryagin's specific water not considered a salt solution for years.

The name "Polywater" appears in the next paper:

E. R. Lippincott, R. W. Stromberg, W. H. Grant, and G. L. Cessac, "Polywater. Vibrational spectra indicate unique stable polymeric structure", *Science*, **164**, 1482-1487 (1969).

Here the authors invent the term "Polywater" for the anomalous water of Deryagin, propose Polywater to be a newly discovered true high polymer of water, and defend the terminology by lengthy speculation. A noncritical reader might be convinced by the degree of speculative structures offered. Infrared and Raman spectra were adduced in support of the invention.

Polywater is made by a recipe using fused quartz or Pyrex capillaries "cleaned by conventional methods and dried"; our students balked at this generality. The Polywater preparations were not characterized by conventional means. Samples were not even weighed; yields could not be given. Occasional spectral data were given in some of the many articles but no satisfactory characterization was ever provided.

The published infrared spectrum of Polywater "is not a spectrum of any known substance". Necessarily, the spectrum of an odd mixture of substances meets this criterion, and our students sometimes made this conclusion. Student concern was greater in that no reference spectrum of pure water itself was presented; only an assertion. Laser probe spectroscopy revealed traces of unnamed cations (sensitivity to 10 pg); copper spark spectra (sensitivity to 1 ng) revealed nothing. Other inadequately described methods revealed oxygen, sodium, and silicon but no halogens; carbon was not a major component. Refractive index was measured and density calculated. The authors assert that purity is of primary importance in distinguishing the properties of Polywater from those of ordinary water but offer no means of telling purity. Polywater was defined operationally, not by physical or chemical properties.

Despite thoroughly inadequate description of the product, the authors proceeded to give a detailed theoretical discussion of likely structure. If DNA structure gained its discoverers such fame, perhaps the Polywater structure will do the same for us. Bond distances in the formal structure were estimated, also the stability of Polywater. In this case, the bond energy value 250-420 kjoule/mole (60-100 kcal/mol) was repeated three times as if to reassure us (see Chapter 4).

This incredible paper by authors from the University of Maryland and the National Bureau of Standards immensely influenced work for the next two years. From Science

Citation Index data showing citations this paper must be very important, despite its error, thus demonstrating that mere numbers of citations does not guarantee acceptance or good quality.

Despite data published in many Polywater papers of the period an adequate chemical description of the material was never provided. Rather, Polywater was defined operationally; you make by recipe, like a chocolate cake: do this, do that, and *voila*. All too often poorly described, operationally defined materials receive major attention in emerging research topics; recall chromatin previously discussed. This problem is still current, witness operationally defined oxidized low density lipoprotein characterized by inadequate physical and chemical properties and bioassay data currently implicated in the etiology of human atherosclerosis.[20]

Careful scientists are accustomed to having new chemicals, special preparations or samples, and microbial agents and newly derived cell cultures meet required formal descriptions prior to publication. Absent such description we have J. H. Hildebrand's comment "Polywater is hard to swallow"[21] and the equation: Polywater = Gollywater = Follywater!

For amusement of our students we assigned a short letter to the journal:

F. J. Donahue, " "Anomolous" Water", *Nature*, **224**, 198 (1969).

This brief letter called to our attention the great danger that Polywater posed to life on Earth. Donahue feared that polymerization of all the water on Earth would now occur. As "Chicken Little" he admonished us to treat Polywater "as the most deadly virus until its safety is established"! The paper drew instant response, was called unduly alarmist and science fiction.[22] One considers that *Nature* published the letter more to stimulate controversy and journal circulation than as a science matter.

Our students were amazed that *Nature* would publish

such a letter. Our own purpose in assigning this paper was to awaken the students once again to the oddities of modern science publishing, both by authors and by journals. Anything in print is better than nothing in print; provocative articles attract subscriptions.

As required of serious science, corrections to the Polywater escapade began:

D. L. Rousseau and S. P. S. Porto, "Polywater: Polymer or Artifact?", *Science*, **167**, 1715-1719 (1970).

In this serious paper the artifact nature of Polywater was advanced but not in certainty. Polywater made by recipe was characterized by the methods previously chosen by others and shown to be "highly contaminated with impurities". Among advanced methods used were neutron activation analyses, X-ray analyses, and spark source mass spectrometry that revealed the presence of Na, Ca, K, Cl, SO_4^- and traces of C, Si, B, N, and O. A call was made for more careful work under well-defined conditions before attributing the "strange liquid" to Polywater, a substance not likely to exist.

Many other Polywater papers were available at this time, some offering additional hypothetical structures, others rejecting the whole business as foolishness. The Rousseau and Porto substantial paper was assigned in our class to acquaint students with what it may take to begin to overcome nonsense that nonetheless receives federal funding.

The matter was further adjusted by:

D. H. Everett, J. M. Haynes, and P. J. McElroy, "Colligative Properties of Anomalous Water", *Nature*, **226**, 1033-1037 (1970).

Further detailed work recognizing Polywater as a hydrosol of silicic acid in water and denying need for postulating the existence of polymeric water was presented. The authors compare infrared and Raman spectra of Polywater with normal water, recognize that the preparations are mixtures, and cite Ockham's Razor in coming to their conclusions.

We could not linger over this paper in detail, as our students were not skilled in the necessary background. However, by this time our students had learned the meaning of title words such as "colligative", but "eutectic" found later in the text still was missed. Ockham's razor also was not a recognized idea among our students.

These last two papers in premier journals served sound notice that the Polywater episode was not a valid phenomenon. Despite their more remote status anent our students' experience the students made a good effort to grasp details and were well aware of the message imparted.

This was a typical PC scandal in that counter arguments were rejected *ad hoc*, and the federal government was deeply involved. Those asking damning questions regarding Polywater existence, characterization, and purity and how explain the stability of a nonexistent substance were drummed from the corps and received no federal money. As for arguments of identity and purity, there were remarks to the effect "Yours may be contaminated, but mine is not!".

The Polywater episode is characterized by: adversary proceedings, federal government intervention and support, *ad hoc* disputation supporting a point opposite to common sense, appeals to morality, sensible controls, reasonableness, etc., aggressive promotions by push-people, no discussion of issues not supporting the agenda, political agendas, provincial discovery and big shot takeover, bandwagon rolling, Nobel Prize suggestions, truth "as I saw it", and suppression of alternative ideas.

Common sense prevailed about Polywater within a few years, such that at an international conference of August 1971 set to cover the topic there were papers supporting Polywater, papers expressing doubts, and papers of retraction.[23] It amuses one to see the arguments for Polywater that were submitted before overwhelmingly convincing evidence was adduced showing Polywater to be a hydrosol containing silica. Characteristic comments among some fifteen papers ranging from support to doubt to rejection are: "two phases or true solution of a macromolecule in water", "results of this laboratory do not refute the existence of Polywater", "presence of some salt is not sufficient to disprove the presence of Water II", "water/anomalous water mixtures formed only half the time", "very unlikely that a polymer of water exists", "reason to seriously question", "strong evidence against the existence", and "Polywater does not exist".

Polywater articles, most published 1969-1972, built to a maximum in 1970, after which abatement occurred. In 1973 Deryagin retracted his work,[24] the episode was over, the world was advised.[25,26] Deryagin resurfaced in the "cold fusion" furor (Chapter 5) with a theoretical explanation of that process.[27]

There is no mention of Polywater in a later serious review of water properties.[28] Nonetheless, water appears to form clusters (trimers, tetramers, pentamers) by weak hydrogen bonding, detected by vibration-rotation tunneling laser spectroscopy.[29]

10. Negative Rate Constants

On occasion we experimented with introduction of other topics of timely interest, mainly to determine how well students could manage topics well outside their expertise but topics of particular interest as regards problems of criticism of the biomedical science literature. One such occasion dealt with the mechanism of ribonuclease action of importance to molecular biology, proposed from peculiar kinetics, kinetics directly criticized by others.

Our selection seemed justifiable from two viewpoints, emphasis on just what labor there is in trying to understand how enzymes function, and, peripherally, just how much trouble one may encounter in trying to adjust perceived error in published papers from laboratories of prominent scientists.

Our experiment did not do well, as the kinetics and mechanism arguments were too far removed from student experience and interest. Once again the depth to which serious work must be conducted to understand enzyme actions offended our students used to descriptive, not quantitative, information. Such papers are simply too detailed, too unnecessarily complex. We have seen this attitude already in the discussion of the review of 3-hydroxy-3-methylglutaryl coenzyme A reductase.

The two papers deal with ribonuclease action from a kinetics approach:

E. Anslyn and R. Breslow, "On the Mechanism of Catalysis by Ribonuclease: Cleavage and Isomerization of the Dinucleotide UpU Catalyzed by Imidazole Buffers", *Journal of the American Chemical Society*, **111**, 4473-4482 (1989).

And

R. Breslow and D-L. Huang, "A Negative Catalytic Term Requires a Common Intermediate in the Imidazole Buffer Catalyzed Cleavage and Rearrangement of Ribonucleotides", *Journal of the American Chemical Society*, **112**, 9621-9623 (1990).

These two papers describing a proposed mechanism of action of ribonuclease were selected both to show problems in details underpinning molecular biology but also to demonstrate other aspects of modern science, of editorial privilege in arranging publication but also of how to get contrary and inconvenient results published.

The proposed mechanism of the two papers involved odd kinetics data leading to the concept of "negative rate constants", a concept devoid of meaning in kinetics despite definition and promotion in the cited papers. Rate constants less than zero cannot be, and in support is the statement "Our kinetic version of it does not seem to be widely known or used".

As was common in our criticism of published work, if glaring problems were not evident we looked for small problems that might suggest error, carelessness, or other peculiarity. One may discount minor editorial matters, such as using the acronym HPLC before its definition on the following page of text. Misuse of the word "since" for "whereas" is a matter frequently observed elsewhere as well. A curvilinear plot in Figure 4B is said to be a Bell curve, but only part of the curve not looking like a bell is drawn.

However, these minor editorial items are accompanied by more serious problems. The expression "5-10 times slower" is used, perhaps meaning one-tenth to one-fifth as fast. Again, this kind of terminology appears in many articles, but such quantitative comparisons of negative properties create uncertainty of meaning; comparisons of positive quantities leave no doubts. Moreover, first order (or pseudo-first order) reaction kinetics were claimed, but ordinates in figures showing rates and in tabulated data have second order rate dimensions μmoles/minute. Thus something is in error.

Attempts by others to correct perceived technical flaws met with rejection of their submissions to the *Journal of the American Chemical Society* in which the original mechanism papers were published. Journal editorial policy precludes publication of corrections unless new experimental data are also provided.[30]

The case illustrates two irregular features of chemistry publication in prestigious journals. One is the privilege extended to prominent chemists (Breslow was president of the American Chemical Society 1996); another is the trick of inserting in proof unapproved addenda regarding matters, material not receiving editorial scrutiny. The correction paper of F. M. Menger has such an addition, he adding that his correction paper was not allowed to be published in the journal having the original Breslow papers in question.[30b] The acrimonious accounts of different referees, taking one side or the other, is reminiscent more of present political unrest in the country than of scholarly discussion of issues. Breslow does not mention the case in his more recent accounts.[31]

11. Matters Arising

Under this topic we undertook to present a matter of increasing concern among scientists, that of science fraud. One recognizes the hoax, misconduct, and fraud as acts of deliberate deception intended to give the perpetrator some modicum of amusement or advantage at the expense of honest folk. I discuss here the several papers we selected for teaching purposes on this matter. A more general discussion of error, hoaxes, misconduct, and fraud is provided in Chapter 6.

The graduate student and junior biomedical investigator must be aware of the insidiousness of science fraud, even if the fraud cannot be recognized at the time. Presumably senior investigators will already be aware of some of these problems through individual experience. These problems range from simple error through several different levels of misconduct to the hoax, to deliberate fraud, and to criminal matters. It is crucially important to distinguish error from fraud, a matter not always possible to achieve given inadequate information at the time. Error is not fraud; fraud is not just error.

One of our most pleasurable classes was that in which we plied the students with several fraudulent papers, papers the students could not possibly determine to be flawed from context, by merely reading the papers, understanding background, or being an expert in the matter. We wickedly chose papers in the reputable journals *Journal of Biological Chemistry, Cell, Nature, Science, Proceedings of the National Academy of Sciences USA*, and *Journal of the American Chemical Society* by prominent scientists from prominent institutions, papers where neither data presented nor arguments advanced revealed fraud. In order to avoid unfairness to the molecular biomedical sciences, we included fraudulent papers on related topics, topics not far removed from what a successful biomedical scientist might want to understand. By taking the students outside their expertise we also forced them to examine some works more thoroughly.

It was our aim to show the very high cost of fraud by requiring the students to write the usual critiques on several papers for one class, a rather heavy load. We usually chose at least one lengthy paper, one that would truly take time to read, let alone study for quality.

We chose to discuss fraudulent papers under the guise of another title "Matters Arising" so that the nature of the papers was not directly disclosed. In our early classes the students had no idea, but in later years some of the students must have been keyed to the fact that we would give such papers for review, and their class critiques were a bit more hedged. Students could not come out and say fraud, except for the one or two notorious ones involving Mark Spector and Efraim Racker.

In order to keep the topic fresh we regularly changed the papers assigned or shuffled the order of our classes so as to disguise matters. Our duplicity was further advanced by including on occasion a sound paper next to a fraudulent one, thereby to confound students even more so but also to emphasize beyond any argument that fraud in science causes great trouble and is to be avoided scrupulously in our own work and exposed where found in work of others.

We distinguished among papers with admitted fraud, with unadmitted but almost certain fraud, with suspected fraud, and with papers retracted or withdrawn for want of reproducibility or for other irregularity. Intensive work by students was necessary to criticize these papers, with our disclosure of their fraudulent nature only after discussing all in class. We declined to grade these written critiques, as generally no student appraisal could have discovered the fraud. Only minor, inconsequential flaws were discovered. Students grasped permanently the high cost in effort of dealing with fraud.

One must approach the question of fraud with care, as deliberate fraud is not just error and error is not fraud. Absent admission of fraud, papers retracted by their authors for want of reproducibility, though suspicious, may reflect error, not fraud. However, there are papers that are manifestly fraudulent whether confessed or no. There are also papers suspected from other evidence to be fraudulent, but in the face of denials and without admissions of fraud there is uncertainty. I present here examples of three categories of papers we regularly used: papers in error, confessed fraudulent papers, and unconfessed fraudulent papers.

Papers in Error. On occasion we presented a case of error resulting in unreproducible work. In such cases, upon discovery of unreproducibility by others or by themselves, the authors withdrew their work, with a simple statement to that effect in the journal. This is the simplest means of adjusting such troubles.

One such example is:

G. Studzinski, Z. S. Brelvi, S. C. Feldman, and R. A. Watt, "Participation of c-*myc* Protein in DNA Synthesis of Human Cells," *Science*, **234**, 467-470 (1986).

The role of nuclear oncogene c-*myc* in cell activities was described as involved in DNA synthesis, a matter of great importance at the time. However, the results were not reproduced,[32] and Studzinski agreed there was error, perhaps because of aging, altered materials.[33] This paper used in class makes a simple point, that of foolish error in using reagents whose identity, stability, and potency are not carefully monitored. Fraud is not suggested in this matter, only avoidable error.

Other cases of self-discovered unreproducible work retracted by authors pose other doubts. Less well defined as to possible cause is the 1983 retraction of two papers of William Thompson dealing with an enzyme acylating a liponucleotide. A contamination possibly present in liponucleotide samples was posed perhaps as cause. Misconduct by Thompson or coauthor G. MacDonald was not suggested, but cause of the problem was otherwise unexplained.[34]

In 1986 two papers of C. Milanese *et al.* dealing with

the ability of lymphokine interleukin-4A to induce interleukin-2 receptors were withdrawn for want of reproducibility. It seems that the work could not be reproduced after Milanese returned to Italy, that the suspect 10-12 kDa interleukin-4A was nonexistent. Milanese admitted his laboratory data were unreproducible, but that pressures in the laboratory had led to the error.[35]

These three unreproducible articles pass from apparent foolish error to suspected misconduct, but each with a common feature. Key biological reagents used in each study were unreliable, being either altered on aging, contaminated with something, or nonexistent. This reliance on biological reagents without very careful verification of identity, potency, and freedom from interfering contaminants remains today a common, serious experimental error.

This problem continues to increase in complexity. One is struck by the many biological reagents required for leadership molecular biology science, example being cited of recent work of Nobel Prize laureates M. S. Brown and J. L. Goldstein in which very complex plasmid constructs are described,[36] with structures confirmed by sequencing all ligation sites. Nonetheless, could undetected error in their syntheses alter results or conclusions.

Retraction of research papers for want of reproducibility by the original authors may be an honorable way to settle matters, but there is usually no acceptable explanation of how the problem came to happen. Was there error or misconduct, proven or suspected. Clearly, with the current litigation rage, honest appraisals by authors may not be possible for fear of lawsuits.

Confessed Fraudulent Papers. We examined cases of confessed and unconfessed fraud. Three examples of each were used in our course at odd times. Among confessed cases are those termed the Gullis Case. Junior investigator R. J. Gullis confessed to frauds in publications from two laboratories, those of C. E. Rowe, University of Birmingham 1973-1976, and of B. Hamprecht, Max-Planck Institute

1975-1976. Work with Rowe involved invented data supporting hypotheses about phospholipid metabolism in brain.[37] Work with Hamprecht also included invented data about cyclic GMP effects on cultured cells.[38] Gullis admitted inventing data supporting hypotheses he believed to be so, and the confessed fraudulent papers were retracted in February 1977.[39]

In our classes we used three of the fraudulent papers, a longer one of Gullis and Rowe or two shorter ones from the Max-Planck Institute as a unit:

R. J. Gullis and C. E. Rowe, "The Stimulation by Transmitter Substances and Putative Transmitter Substances of the Net Activity of Phospholipase A2 of Synaptic Membranes of Cortex of Guinea-Pig Brain", *Biochemical Journal*, **148**, 197-208 (1975).

This longer paper with confessed fraudulent data was a bit too much of a burden for one class meeting and was not regularly assigned. It would not be easy to determine that the paper was fraudulent from its presentation. One clue would be "Most experiments were repeated at least once".

Two other papers with related confessions of fraud were more often assigned:

M. Brandt, R. J. Gullis, K. Fischer, B. Hamprecht, L. Moroder, and E. Wunsch, "Enkephalin regulates the levels of cyclic nucleotides in neuroblastoma x glioma hybrid cells", *Nature*, **262**, 311-313 (1976).

And

R. J. Gullis, J. Traber, and B. Hamprecht, "Morphine elevates levels of cyclic GMP in a neuroblastoma X glioma hydrid cell line", *Nature, 256,* 57-59 (1975).

These papers attempt to demonstrate that morphine and endogenous pentapeptides with opiate-like activities act by regulating levels of cyclic nucleotides in brain. Again, in neither case is there obvious means of detecting error or fraud. Only a glaring editorial blemish occurs, where the ∞ symbol (infinity) is used in figure abscissas for extreme values.

Another confessed case is the Rosner Case:

M. H. Rosner, R. J. De Santo, H. Arnheiter, and L. M. Staudt, "Oct-3 Is a Maternal Factor Required for the First Mouse Embryonic Division", *Cell* **64**, 1103-1110 (1991).

The paper has an assertive sentence title alerting the critical reader to possible lapses in judgement or achievement. Other alerting hints include undefined terminology known only to those closely acquainted with the topic: acronym POU for Pituitary Oct Uterus, Oct-3 as a transcription factor expressed in mouse oocytes, GAPDH (glyceraldehyde phosphate dehydrogenase), POUBOX, and OUGO. Besides these blemishes, there is no obvious means of determining fraud from reading the paper. Rosner admitted that he fabricated data. The paper was retracted.[40]

The third confessed case is the Zadel Case, involving the great present interest in the resolution of synthetic racemic compounds having important biological activities. Much current commercial activity centers on one or the other enantiomer as an active agent, and methods for such resolutions are in demand.

Thus, two papers dealing with resolutions of enantiomers published in the same journal issue in 1994 are cited here as good examples for student instruction in what can be done with creative thinking.

G. Zadel, C. Eisenbraun, G.-J. Wolff, and E. Breitmaier, "Enantioselective Reactions in a Static Magnetic Field", *Angewandte Chemie International Edition*, **33**, 454-456 (1994).

In the Zadel case there was claim of an absolute asymmetric synthesis, or enantioselective synthesis, from achiral substrates without the presence of chiral agents but by use of a static magnetic field![41] Despite theoretical argument that such event be impossible[42] the appeal of the possibility of resolution of enantiomers from achiral substrates under the influence of a magnetic field was great. Had these results been genuine there would have been a major improvement in understanding the origin of single stereochemistry (L-amino acids, D-sugars) implicated in life's origin.

Other than the possible theoretic objection there was no hint of irregularity or misconduct in the paper. However, these results could not be reproduced by others; indeed positive results were obtained only when Zadel was present. Zadel subsequently confessed to fraud. It seems he added resolved product to the reaction system in such a way as to obtain enantiomeric excess in the final analyzed product. The fraud was discovered within four months and the work disowned.[43]

For contrast, on occasion we assigned sound papers together with flawed or fraudulent ones, thereby to keep the students on their toes. Thus, we assigned a companion paper that also dealt with new methods for resolutions of racemic mixtures.

J.-L. Reymond, J.-L. Reber, and R. A. Lerner, "Enantioselective Multigram-Scale Synthesis with a Catalytic Antibody", *Angewandte Chemie International Edition*, **33**, 475-477 (1994).

In this case, we assigned the second paper published in the same journal issue in which a catalytic antibody (abzyme) was used to resolve gram amounts of material. Our students did not recognize the term abzyme, nor did this paper seem important to them.

In neither paper would a reader have detected misconduct or fraud, indeed any serious flaw or error. The two papers together made a good pair for criticism in our course, one being admitted fraud, the other revealing new approaches toward important goals of synthesis. Use of these chemistry papers took our students well outside their experience. The topic is, however, of growing importance away from molecular biology.

The crucial importance of optical antipodes of known racemic drugs is now under commercial exploitation. Popular drugs, such as Prozac (racemic fluoxetine) for depression, are now being resolved; one active enantiomer is to be readied for regulatory approval and for the market. Resolution of intermediates needed for the synthesis of epothilone anticancer drug prospects using catalytic antibodies has been accomplished, making catalytic antibodies of greater versatility and importance in syntheses and possibly production as well.[44] There are now annual professional meetings Chiral USA devoted to such matters.

Unconfessed Fraudulent Papers. Papers acknowledged to be frauds by some authors but not by all, particularly the miscreant perpetrating the fraud, may be retracted or withdrawn without a confession. Three cases exemplify the circumstance. The first is that of Ph.D. candidate Mark Spector, another that of post-doctoral associate Monica P. Mehta. Both cases involved junior associates of internationally prominent investigators who may have been a bit too enthusiastic over new results and too busy with their overall interests.

In the Spector Case we have one of the saddest examples of discovered but unconfessed fraud, that of Mark Spector working with Efraim Racker. We regularly assigned one of the notorious fraudulent papers of Spector, although we suspected that later students were aware of the matter.

> M. Spector, S. O'Neal, and E. Racker, "Regulation of Phosphorylation of the ß-Subunit of the Ehrlich Ascites Tumor Na^+K^+-ATPase by a Protein Kinase Cascade", *Journal of Biological Chemi*stry, **256**, 4219-4227 (1981).

Spector proposed a cascade of protein tyrosine kinases controlled phosphorylation of the Na^+K^+-ATPase of cultured Ehrlich ascites tumor cells. He became an instant celebrity and was invited to speak at the Gordon Research Conference and to give other presentations.

Lionized though he be, the work was suspect. Only Spector could get the results. Others in Racker's laboratory attempting to check the work were frustrated.

The fraud was ultimately disclosed accidentally when an

analysis of the radiation emitted by Spector's samples was identified as that of γ-emitter ^{125}I of iodinated protein and not that of ß-emitter ^{32}P of the $[^{32}P]$-ATP supposedly used. Several other papers are also suspect.[45] The fraud remains unconfessed, but Spector did not get his Ph.D. degree from Cornell University and has disappeared from biomedical science. In 1983-1984 Racker spent eight months at the bench, he trying to determine the true facts of Spector's work, to no avail.[46] Racker has since died.

Another notorious unconfessed case is the Mehta Case:

R. Breslow and M. P. Mehta, "Catalytic Directed Steroid Chlorinations with Billionfold Turnovers", *Journal of the American Chemical Society*, **108**, 2485-2486 (1986).

Three 1986 rapid communications of Breslow and Mehta deal with brief announcements of novel chemical functionalization of steroids mimicking enzyme transformations.[47] We assigned only the first of these three communications, as the chemistry was beyond our students ready comprehension. These three papers were withdrawn by Breslow, stating that some of the work could not be confirmed.[48]

These papers continue Breslow's work on organic chemistry mimicking enzyme actions. By use of a steroid substrate and a special reagent reaction at otherwise unaccessible sites of the steroid structure was claimed with billionfold turnover rates. The synthesis could have lead to intermediates potentially yielding commercially important steroid hormones, thus was of importance from biomedical and business aspects.

Such high turnover numbers are exceptional, and confirmation by others of the laboratory was sought. Mehta's results were initially confirmed, but her exclusion from the laboratory and repetition of the work "under secure conditions" resulted in nonconfirmation. She has not confessed to misconduct.

The earlier Axelrod Case involves retraction by two

authors but with no confession of fraud by one of the authors:

L. R. Axelrod, P. N. Rao, and D. H. Baeder, "A Steroidal Analgesic", *Journal of the American Chemical Society*, **88**, 856-857 (1966).

And

L. R. Axelrod and D. H. Baeder, "A Steroidal Analgesic", *Proceedings of the Society for Experimental Biology*, **121**, 1184-1187 (1966).

The 1966 communication describing a steroid 2,3,4-trimethoxyestra-1,3,5(10)-trien-17ß-ol with analgesic activity more potent than morphine (!) drew instant attention from the pharmaceutical industry. Were this observation valid the authors would become immediate celebrities. However, the bioassay data could not be reproduced by others, and the papers were withdrawn.[49]

The synthesis work remains correct; the bioassay data conducted by Baeder at Mallinckrodt Chemical Works, St. Louis, MO, appears unsound and possibly to be fraudulent. It defies common sense to make claims of important physiological activity and not understand that they will be instantly checked and exploitation of any true leads followed vigorously. Knowing the market for steroid drugs, any report of new steroid activities brings instantaneous examination by potential competitors.

We assigned the two papers for criticism, but had only limited success in instruction, as our students had so little background in chemistry and pharmacology. Also, only minor flaws are apparent; one steroid was named as an ether where systematic nomenclature would name it as an ester. In the synthesis paper no range of effective doses was provided, only reference to the bioassay paper in press. The obsolete term mµ for wavelength was used, hardly cause for alarm.

There are numerous minor flaws in the bioassay paper: $LD_{50}s$ was used, presumably the plural of LD_{50}; jargon details included use of a Sanborn model 350 device for measuring blood pressure and Lead II, Physiograph, terms perhaps recognized by physiologists but otherwise incomprehensible. There were undefined acronyms PSP, SGOT, SGPT, EKG, ACTH and MP-2001 code number for the test steroid, mixed usage kg and kilo for kilogram, and use of the indefinite pronoun "they". From references to "personal communication" it appears that the steroid was administered in doses up to 1 mg/kg to volunteer human patients for control of post-operative pain or chronic pain associated with malignant tumors!

12. Student Compositions

As Seneca said long ago *Non est ad astra Mollis e terris via* (There is no easy way from earth to the stars).

In the second part of the course we exposed the students to how it feels to have your own papers criticized right there in class by others intent on discovery of deficiency and error. We had students compose a ten-page review of a topic selected from lists of acceptable topics, after which the same or modified topic was to be used to prepare a ten-page model grant application for research funding.

It is *de rigueur* in many graduate programs to require students to write formal grant requests, either as learning experiences or perhaps to supply faculty with added resources for their own funding attempts. Such grant writing precedes a normal sequence of composition of thesis or dissertation, abstract writing, and writing experimental and other sections of a research paper. I did not agree that grant writing was of such immediate importance, but our course flourished with those arrangements nonetheless.

We both prepared for class a list of topics from which students selected their fate. Choice of a subject was made after a chance draw of a number, the order in which selection was to be made. Only one topic could be presented; if your favorite was taken before your turn to chose, pick another. The topic selected had to be remote from the student's immediate graduate research project, so each was on relatively new material according relatively equal bases. In a few cases attempts were made to circumvent this restriction, with an attendant loss of credit.

Necessarily, the topics listed were within our individual interest and expertise. The two lists emphasized our specialty interests in molecular and cell biology on the one hand, in biochemistry on the other. It was instructive to see the titles selected by the students for their reviews and subsequent research proposals, also to review students' progress in tending to detail in their research proposals, given heavy criticism of the initial review papers.

Specific titles of several biology review papers and derivative research proposals from our most recent classes include: (1) Mechanism and effects of fos oncoprotein phosphorylation and proto-oncogenes c-*fos* and c-*jun*. (2) Recessive oncogenes and the retinoblastoma locus. (3) Genomic imprinting and DNA methylation. (4) G-Proteins: Characteristics and mechanisms of interaction with adenyl cyclase. (5) Protooncogene c-*myc* and its expression regulated by hormones. (6) Skeletal muscle myogenesis with special emphasis on the role of the MyoD1 gene. (7) Unveiling the mysteries of cystic fibrosis. (8) Regulation of splicing in messenger RNA precursors. (9) Negative transcriptional regulation of eukaryotic gene expression. (10) Olfactory signal recognition and transduction. (11) Regulation of vitellogenin gene transcription. (12) Nerve growth factor and its functional association with health and disease. (13) Immunocytokines in the nervous system. (14) Apoptosis and the Bcl-2 family and Bax function.

These topics range through matters of current appeal for the biology students. Although our students on balance were far more oriented towards molecular and cell biology and so chose their subjects, some chose biochemistry topics such as: (1) *In vivo* oxygen metabolism. (2) NMR in Cell Biology. (3) Biochemistry of Vitamin D. (4) Vinblastine and vincristine: Medicinal products from the Madagascar periwinkle. (5) Oxygen toxicity on the skeletal proteins of red blood cell membrane. (6) Application of ^{19}F NMR in ligand-macromolecule interactions. (7) ^{113}Cd as a structural probe in NMR studies of biological metal coordination sites. (8) The function and regulation of phosphatidylinositol 3-kinase. (9) Oxygen species and aging.

Here the expertise of the instructors was of paramount importance. Clearly, we had to be well ahead of the students with respect to current literature. No student could write a paper that would not be graded by an expert on the topic. Nonetheless, on occasion I and my biology colleague learned some things we had not noted previously.

Students were advised that their papers would now be graded for content and style and marked down for poor English grammar and syntax. Although Orwell's six concise rules for English compositions[50] were not explicitly emphasized, their simple meanings were conveyed in class:.

1. Never use a metaphor, simile, or other figure of speech which you are used to seeing in print.
2. Never use a long word where a short one will do.
3. If it is possible to cut a word out, cut it out.
4. Never use the passive where you can use the active.
5. Never use a foreign phrase, a scientific word, or a jargon word if you can think of an everyday English equivalent.
6. Break any of these rules sooner than say anything outright barbarous.

Note that these rules have been disobeyed in the present screed in many places. Necessarily, scientific terms have had to be used, and the passive voice has been inculcated into most scientists so that this breach of writing occurs also.

Student presentations were assigned by rota, the first two students to present their review papers then being the last to present their research proposals. Usually student presentations involved a brief 20 minutes oral presentation of their review or proposal, followed by questioning by the other students. Initial oral presentations by the first two students were frequently flawed in several ways. Chief among these were presentations given sitting rather than standing, informal everyday clothing instead of a bit more formal wear, slouching or leaning on the table instead of standing erectly and taking charge, use of overhead projector inefficiently by turning it on and off repeatedly, and perhaps drinking from a soda pop can. "You know", "Uh", "Well", and other interjections regularly appeared.

Inasmuch as we remarked on these offenses and marked them down in student grades an immediate improvement resulted. One could see correction of these faults in the next set of presentations, as the students rapidly learned the tricks of oral presentations. To inform, to amuse, to convince others one must have a professional appearance. Students generally were unaware of this need until they were so informed, either in our class or, depending on who supervised seminars, the matter was discussed there.

I had supervised our department seminar one year, and I made notes of criticism for each seminar presentation and sent the highly critical notes privately to each student. A good performance was lauded; a poor performance was severely criticized. Other faculty liked to give grades for seminar performance, but I preferred to analyze presentations and inform the student privately. Students told me that they greatly appreciated the private criticisms; it would have been devastating for such remarks to be passed in the seminar classroom with all to hear.

In those presentations of research proposals for our class, as well as for formal presentations by graduate students to fulfill their obligations for candidacy, following general discussions and questioning I liked to ask "Where will this proposal fail"? Students that had not been primed for such a question were frequently at a total loss for words, as in their enthusiasm they had not given this possibility much thought, if any thought.

After each student presentation, whether review or research proposal, we gave the student a written critique and number grade on three aspects: background and content, written style, and oral style. These papers had to be typed and were marked down for poor English, bad grammar and syntax, poor organization, use of undefined acronyms and jargon, and the usual puffery of indoctrinated students on a topic of their love. Most written work passed our scrutiny well enough, but there were a few outstanding failures to perform.

In one case a student copied too many passages from a source and was accused of plagiarism by another student. In another case, a research proposal was taken from one that the student's mentor had written. In yet another, a complete disconnect between papers cited in the text and the reference list occurred; clearly the student had not done a proper editing job. These students had their grades reduced.

Most research proposals and reviews were competent, although the naïvité of our students was generally revealed in this phase. The frequent reach-exceeding-grasp phenomenon was common among our students, who conceived grandiose problems but had difficulty designing solutions.

CHAPTER 3. SOME CURRENT PROBLEMS

Wer denkt, zweifelt schon (Whoever thinks already doubts)
- Cornelia Berning

Among current problems of science is the question of where science is today. There are those who pronounce that everything of fundamental importance in science has already been discovered; we know about the electron. Before World War II it was said that physics had discovered all that mattered, radioactivity, television, etc., and needed only to refine the decimal points.

We see titles of articles and books that presage the doom of science: "The Twilight of Integrity", "The Twilight of Science", "Heyday of Science Seen as Over", "Are Researchers Trustworthy?", "Cheating in Science", "Is Science Really a Pack of Lies?", and "The End of Science". The scholar shall not be deterred, although the science impresario may be daunted.[1]

The problems of current biomedical research are not all technical ones, as we now live with considerable distrust of science despite the obvious advances in human society given us by modern science. We regularly see mention in the public domain of "creation science", social science, and even of scientific wrestling! Engineering and medical practices are routinely termed science. These and other peculiar uses of words for reasons other than honest communication are given special treatment hereinafter under the guise Beware the Meaning of Words in Chapter 4.

Moreover, several other matters continue to influence adversely the proper advance of science. Among such are the anti-science remarks of prominent critics, news media hyperactivity into the blue skies, disputation over religion and ethics of experimentation, and the overwhelming problems of funding research. The general anti-intellectual atmosphere felt by many scholars is particularly evident in the realm of science, where the mad scientist or simple fool is a standard character in motion pictures and television programs. A few specific arguments exemplify the present status:

1. Attacks on Science and Sensibilities (Anything But Knowledge)

One of the most amusing attacks on science is that by "creation science" and "intelligent design" advocates who seek to discredit Darwinian evolution in favor "equal time" in science classrooms for study of religious disputation about human origins. "Intelligent design" poses the notion that an undescribed "intelligence" designed the universe, as our present universe is too complex to be a product of chance. These movements use the same techniques of the Polywater (Chapter 2) and "cold fusion" (Chapter 5) episodes: press releases, national conferences, and advocacy books but no published peer-reviewed journal articles.[2]

Attacks on science are not solely those of creation scientists or advocates of "intelligent design". Some with no pretense at all to science set the pace. From the educator John Dewey (1859-1952) we have such leadership remarks: "It is one of the great mistakes of education to make reading and writing constitute the bulk of the school work in the first two years", "The mere absorbing of facts and truths is so excessively individual an affair that it tends to pass into selfishness. There is no obvious social motive for the acquirement of mere learning, there is no clear social gain", and "Children who know how to think for themselves spoil the harmony of collective society which is coming, where everyone is interdependent".

The emphasis on social gain clearly disadvantages scholarly achievement of all kinds and science greatly. Serving social gain smacks of serving the state; paraphrasing

Heinrich Himmler *ca.* 1933: "Germans who wish to use firearms should join the SS or the SA - ordinary citizens don't need guns, as their having guns doesn't serve the state". This remark reveals motive behind some attacks on science, science that is not controlled by the state, by those seeking unlimited power over society, thereby to make it "nearer to the heart's desire".

Yet other influential writers stress strange attitudes about science. Poet-classicist Anne Carson, McGill University, Montreal, PQ, opines: "Scientists operate under the 'happy delusion that there are such things as facts'".[3] As developed hereinafter, we see foolishness turn to wickedness in PC now prevalent in the land. One sees such cruel quotations as: "Our aim is simply to encourage sensitivity to usages that may be imprecise, misleading, and needlessly offensive", but then capped by "Man, like other mammals, breast feeds his young".[4] It is PC that is imprecise by design, deliberately misleading, and needlessly offensive to educated persons.

Anti-science activists have much to say in the political arena against science. Indeed, there is a strong movement against science as scientists have built it. Science is viewed by social constructivists as being socially constructed rather than as being a reflection of the real world. We see regularly the rise of pseudoscience and "New Age" culture. One author remarked about his work "This book is dedicated to all the science teachers I never had. It could only have been written without them".[5] Even the treatment of chaos theory by nonscientists is inimical of science.[6] The purveyors of PC, the so-called academic left, find fault with the concepts of science.[7] Other PC and feminist anti-science attacks include rejections of "Eurocentric" science, science conducted by dead white men, thus with the same arguments also made against poets, authors, and political leaders. We see the notions as deconstruction (discredited word-built value systems), postmodernism (all values and knowledge are socially constructed), and feminism in the remarks "corrupt interior of science", "built on oppression of others", and "call for female-friendly science".[8]

Other disciplines as well as science are subject to these influences. Revisionist historians at the Smithsonian Institute, Washington City,DC, proposed that ending World War II with the nuclear bomb dropped on Japan by the B29 Enola Gay was all wrong. Also, a Smithsonian Institute display about modern chemistry emphasized all the environment problems of chemicals without mention of the simple fact that everything is made of chemicals. Should inconvenient facts bother a government agency?

Such notions are not dangerous or distressing except to academic scientists who may more fully seek the truth of a point. In the guise of science meetings on current political issues such as that of the New York Academy of Science (June 1995) have been arranged,[9] and oodles of books are now emerging.[10]

Examination of race and sex differences receives extraordinary PC scorn. Offensive words must be changed; conclusions of serious studies must be judged according to whether a "hate crime" has been committed. Angry protests arising over such books as The Bell Curve: Intelligence and Class Structure in American Life by R. J. Herrnstein and C. Murray and J. P. Rushton's Race, Evolution and Behavior are instructive of the extent PC has invaded science.[11,12]

Among other problems of the public view of science is that of "junk science" developing in law courts. As with the pejorative "junk guns" for inexpensive firearms, so we now have "junk science" testimony of experts based on personal experience, informed or uninformed, opinion and belief, testimony that cannot be understood by judges, juries, lawyers, and nonscientists.

A recent Supreme Court decision[13] sets guidelines for determining what science evidence may be admitted to law court; judges are given discretion to decide what science matters may be introduced. The difficulty can be grasped by court findings involving silicone breast implants for women, fluoridation of drinking water, and the recent flap over plasticizer phthalates in plastic toys. Here pursuit of stipendiary interests prevail; science is not the issue at all.

2. Dumbing Down

This problem starts in early public school days, where "dumbing down" is now necessary to meet government and PC notions. We see multiculturism, racial diversity, feminist concepts, and environmentalism promoted even in exact sciences, "fuzzy math" texts with President William Jefferson Clinton's picture and Maya Angelou's poetry.[14] Students are encouraged to work together rather than individually, to arrive at group consensus as to answers, with a good explanation of how a wrong answer was obtained getting the same credit as a correct answer, this to a simple direct mathematics problem.

Such deep insight into mathematics is revealed in the statement "what good is it to solve an equation if it is the wrong equation?". By whose judgement?

There are other aspects of the dumbing down. Civil rights and PC advocates object to standardized tests of academic achievement as being discriminatory. What else should a test be? The latest flap over college entrance examinations requires that discriminatory questions be deleted, thus that all classes of students make the same test scores. So, a question about snow is unnecessarily discriminatory and must not be used, as southern Negroes are deemed not to know what a snow bank is!

Those unaware of or denying the decline in public education need only examine the McGuffy readers of the early twentieth century to recover their modesty. However, another factor must enter into consideration, that of increased costs of public education for students with "learning disabilities". As increased funding for such students is discovered, their number increases, the number of teachers dealing with those students increases, and administration costs increase.[15]

3. Education versus Training

In our modern age we see strange notions creeping into what used to be the realm of scholarship. We see regularly such confusing terms as "teacher training" but also "driver education", thus that teachers should be trained rather than educated, that drivers be educated rather than trained to manipulate their automobiles. Teachers have become "facilitators"; students are "clients". Self-esteem, success, and accomplishment are more important than learning.

The practicing physician must be trained in the medical arts, but as training in thinking does not work, the physician must also be educated in the medical sciences. Thus, the medical school must provide both advantages. The science student in graduate schools associated with medical schools should be educated in the biomedical sciences as the major objective, as training to operate this instrument or that procedure cannot sustain a career most likely to change drastically, given modern progress in science.

However, in the biomedical sciences the continuing confusion between education and training may be deliberate, as pressures to have trained hands but not educated collaborators are strong. We have NIH "training grants" for science graduate students to provide hands for funded investigators who will not allow their graduate student trained hands to take elective courses in subjects not absolutely necessary for the hands' minimum degree requirements.

The long accepted functions of medical schools and great universities to teach and to think[16] are being subverted. The university itself is to be an instrument for social change, not for scholarship. The research university and medical school must be devoted to the business of fundable research, the attraction of massive federal funding being what it is.[17]

The concept of scientist and physician as scholar versus

as highly "trained" technician must be carefully considered. Training tends to limit what one can achieve in life to the trade in which one is trained. By contrast, they being trained for nothing feel free to attempt anything. Poet Laureate John Masefield aptly puts it: "But trained men's minds are spread so thin; They let all sorts of darkness in".

However, social problems are causing medical education to be in a state of change. Besides the usual arguments for more diverse medical student populations, proposals to eliminate organic chemistry as prerequisite for medical school admissions now have surfaced. It is as if formal biochemistry that requires basic organic chemistry for proper understanding is no longer needed. New curricula just go directly to descriptive molecular biology without the science background. Students without organic chemistry are plunged into molecular biology charts with organic chemical structures showing hydrogen bonds, just about as understandable as Mesopotamian cuneiform writing.

Pig-Latin is easily mastered once the realization that it be Pig-Latin being spoken. The comedians' "doubletalk" of yore is understandable once it be realized that nonsense babble is involved. Even the "Upper Language" of comedian Victor Borge (born Boerge Rosenbaum, 1909-2000) is acceptable, once you have the key. Raise every word or syllable that sounds like a number by one; thus, Borge's song "My too too wonderful lieutenant" became "My three three twoterful lieutelevenant". Borge also invented oral punctuation of speech, using peculiar sounds from the mouth for comma, semi-colon, colon, period, question mark, and exclamation mark, all the more allowing further finer distinctions of meaning in speech.

Premed students had to have comparative anatomy and quantitative analytical chemistry (including limestone analysis!) when I was an undergraduate 1944-1946. Both requirements were long ago dropped. Now organic chemistry may be the next preparatory course to be dropped, as no longer relevant in our molecular world.

Medical students at one time spent much time at the

bench studying stool. Now that "molecular" processes have triumphed that aspect has disappeared from the curriculum; people no longer get worms anyway.

Very recently another of the periodic "new curriculum" based on problem solving by small groups of medical students guided by a facilitator has been introduced in our medical school. Medical problems are outlined, and the unaware medical student must determine what is needed to diagnose and remedy the case, this without having adequate didactic course work to acquaint the student with what is already known about the applied aspects of basic knowledge. Ask about; somebody may know; go to the library; check the computer; get "on-line".

4. Dogma

Not in the same vein but impediment to science advances nonetheless is dogma, whether official or otherwise. The Roman Church veneration of the Shroud of Turin discussed later in Chapter 5 is one sort, but unofficial dogma and beliefs among genuine scientists also recur in various modes that may impede science. Chemist Emil Fischer believed that organic compounds with molecular weights greater than 5,000 could not exist, thus in essence denying the possibility that high mass polymers could exist! Rather than long chain polymers, physical aggregates were posed.[18] Only after the 1920 alternative proposal of Hermann Staudinger, Nobel Prize in Chemistry 1953, that high mass polymers be repeating monomer units covalently bonded was the matter resolved. The subsequent development of synthetic polymers so pervasive in our current lives and the emergence of natural biopolymers of vital importance to living systems make it almost impossible to grasp the effect this dogma had on progress in the period 1913-1926.

Three dogmas involving steroids encountered by the author typify the effects such thoughts may have on the progression of science. Prior to the burst of synthesis activity of the 1950s period that gave us the vast array of new corticosteroid analogs so beneficial in medicine there was a common dogma that improvement over the natural hormones hydrocortisone and cortisone was not possible. It was possible, of course, and the dogma was shortly thereafter overthrown with the accidental discovery of more active synthetic analogs prednisone and prednisolone. A flood of even more highly effective synthetic analogs followed.

A similar in-house dogma that the acetonide derivative of the synthetic corticosteroid analog triamcinolone would not be active blocked testing at both Lederle Laboratories, Pearl River, NY, and Squibb Institute, New Brunswick,NJ. Both companies were then producing triamcinolone. At Lederle no tests of triamcinolone acetonide were conducted at all until after the Squibb announcement of its activity, discovered accidently. It had been thought at Squibb that unwanted traces of the physiologically active steroid precursor 9α-fluorohydrocortisone and byproduct 9α-fluoroprednisolone might be present in finished triamcinolone product. Thus, bioassay for these bioactive products was sought by forming the acetonide of finished triamcinolone, the acetonide then deemed to be inactive, again without testing triamcinolone acetonide itself for activity. By this ploy the high activity of triamcinolone acetonide now long a commercial product was eventually discovered 1956-1957.

The dogma that synthetic steroid hormone analogs be inactive was shattered forever. This error is no longer tolerated. Bioassay everything!

A more recent dogma surrounds study of the mode of action of steroid hormones and drug analogs. Early hypotheses posited now abandoned notions of steroid interaction with enzymes as coenzyme or as allosteric modifiers. The concept of steroids as regulators of gene activity followed demonstration by Peter Karlson of steroid action on insect salivary gland giant chromosomes. Note the fundamental discovery and explanation was not with steroids of current biomedical interest but with the arthropod molting hormone ecdysone, once again evincing the basic science unrelated to funded mission-oriented research.[19]

With the discovery of nuclear steroid receptors 1960-1970, their biochemical characterization 1970-1985, and their cloning with molecular biology methods 1986-1990 there developed the dogma that *all* steroid hormone action was moderated by nuclear receptors that are transcription factors. The dogma arose and persisted despite the early 1941 discovery of the rapid anesthetic action in rats of intraperitoneal administration of progesterone and much more recent information of rapid extragenomic steroid effects.[20]

Clearly the concept of dogma exists among "molecular" scientists as well. Mantras "One gene, one protein" and "DNA begets RNA begets protein" are now under reconsideration.

5. There Was A Time

As already noted, there was a time when we got a grant to do research. Now we do research to get a grant. Also, there was a time when a scientist's remarks were considered either true or false. Now, like remarks of politicians and business persons, one must ask "Why is he telling me this?". These two damning comments on today's biomedical research atmosphere necessarily fail to make similar statements about the past when other circumstances were the case: few funds, inadequate facilities, but freedom to do whatever; explored from a fundamental viewpoint, with commercial exploitation conducted by industry and not faculty members; and nonetheless, the potential for error, misconduct, and fraud now of concern.

There was also a time not too many years ago when our department faculty represented the several major subjects traditionally included in medical education. However, over the past three decades this status has changed. Generalists or traditionalists skilled in their topic but also devoted to academic matters, including teaching, have been displaced by specialists (read this to be funded researchers). As the lure for big federal research grants has seduced administrators, so has the composition of faculty been altered. "Molecular" faculty now predominate; that is where the money is.

Traditional faculty were generally recruited and retained so that the several topics of importance were covered by an appropriate faculty member. Some gaps in coverage occurred at times, but effective coverage was arranged somehow. This mode of operation began to change about 1970, with reorganizations of departments and emphasis on specialization, building research empires. There must be a "critical mass" in order to build sub-institutions, to attract generous federal research funding.

As traditional faculty retire or resign they are replaced by persons with specialty research interests compatible with those of the sub-institutions. More is merrier. However, it became clear that the funded researchers were not necessarily able to provide sound teaching to medical students. We had faculty who could not work pH problems, so acid-base balance was simply dropped from our medical biochemistry teaching to accommodate! Other newly hired, successfully funded faculty proudly announced that we did not need a broad spectrum of graduate school faculty interests, that one individual faculty member could teach the graduate students "all they need to know"! Molecular knowledge, of course.

A related phenomenon is occurring in faculty recruitment. Entry-level appointments are not being offered, at least where tenure may be gained eventually. Rather, more senior, funded faculty are recruited, with expectations of continued success in arranging generous extramural funding for the institution. Those junior faculty actually appointed on the "nontenure" track are vulnerable to dismissal if their funding efforts are unsuccessful. Thus flourishes administration!

The present political furor aside, we are inundated with

problems of which account to accept as valid, which be useful to our thoughts. In the entertainment/news business celebrity newscasters adjust their material incessantly to suit their needs to remain popular, to sell their products, and to advance their political agendas. Their blatant "spins" are there to see every day. The compulsion to be a celebrity with increased income overcomes fairness, good judgement, and honesty, witness recent (1997-1998) faked stories about heroin use in London on CBS "60 Minutes" entertainment television show and about the "Operation Tailwind" nerve gas attack episode of the Vietnam war as told in *Time* magazine by CNN celebrity Peter Arnett and April Oliver. Fabricated stories by journalists Stephen Glass of the *New Republic* magazine and Patricia Smith of the *Boston Globe* newspaper turned each into a celebrity until exposure. Smith was a finalist for the 1998 Pulitzer Prize!

At issue is whether celebrity status of biomedical scientists is not now similarly out of control. As the rapacious scramble for big federal research funds by academic institutions makes fund raisers of their faculty, celebrity status becomes necessary for some to survive. Celebrities are excused from the ordinary work of educational institutions, as their money-raising is so much more important, and noncelebrity faculty can do the teaching anyway. The transformation of medical school faculty from teaching and research to funding agency has now occurred. Sadly, the scholar scientist is being displaced by the funded science salesman.

The classic notion of the scholar and scientist seeking to understand Nature and the world may not have passed entirely, as there are still those who behave in just this fashion. However, given the stakes of high finance and the goodies therefrom, personal behavior approaches that of business. In order to pay oneself a handsome stipend and to have a telephone with several buttons for different lines, the best desktop computers, a private secretary, foreign travel funds, and somewhere down the line funds for research personnel and laboratory supplies and activities, one must have a generous federal research grant in support. Without the grant the scholar is almost unwelcome in biomedical institutions. The motivation of biomedical scientists must now be evaluated.

6. Motivation

These points raise another issue, that of just what is a scientist anyway. The issue of training versus education already discussed is paramount, and the seduced entry into study of science may not be all that suitable for many a student, given other opportunities.

Then there is the matter of motivation. William Osler asks "Who can understand another man's motives? Does he always understand his own?".[21] Those with genuine callings need no guidance, surely no impediments to their progress. Less motivated students subject to persuasion may not be helped by present methods of NIH funded "training".

One of the philosophical notions advanced on occasion is that pursuit of rewards in science may be poor reasons for becoming a scientist, even a molecular one. The Nobel Prize is ever before us, and "molecular" scientists are winning their share. There are other prominent awards for achievement, but the Nobel Prize is the one many aspiring scientists see themselves winning, and not too far in the future. Moreover, the Nobel Prize was awarded, at least once, for error! The 1927 award for chemistry to Heinrich Wieland was for his work with the steroid bile acids, with his assigned structure in error. Remember this was 1927.

However, we must recall Alfred Nobel's basis for awarding the Nobel Prize, "to those who, during the preceding year, shall have conferred the greatest benefit on mankind", thus an altruistic award for great public service.

Instant fame and fortune is reflected in the Nobel Prize award to James D. Watson and Francis Crick 1962 for their description of the double helix structure of DNA.[22] Watson had been a Quiz Kid on the old radio show. While enjoying a fellowship in the Cavendish Laboratory, Cambridge University, he and Crick were given access by Maurice Wilkins to X-ray data of Rosalind Franklin, from which data Watson and Crick devised their double helix structure. Franklin, who had previously suggested a helical structure from her data, did not know or approve of the disclosure.

The propriety of surreptitious use of others' data without their knowledge or approval did not enter the picture of the 1962 Nobel Prize awards to Watson, Crick, and Wilkins. Franklin had died in 1958, thus averting possible conflicts. Thereby are sown the seeds of instant success - a stroke of genius but not of careful laboratory work by the Nobel Prize laureates.

Others have been motivated accordingly, and more than one case of rivalry has occurred. The 1977 Nobel Prize in Medicine awarded Andrew Schally, Veterans Administration Hospital, New Orleans,LA, and Roger Guillemin, Salk Institute, LaJolla,CA, for their independent work on hypothalamic releasing hormones is a prize example in the lack of collegiality and cooperation among contenders.[23] The more recent controversy over priority of discovery of human immunodeficiency virus (HIV) infection leading to AIDS pits Luc Montagnier of France and Robert C. Gallo, NIH. Disputed priority is not a new matter but has existed from the very early years of modern science, witness Isaac Newton versus Gottfried Wilhelm Leibniz, Charles Darwin versus Alfred Russel Wallace.[24]

Others gain the Nobel Prize by other means. Rigoberta Menchú, 1992 Nobel Laureate for Peace and activist for human rights of Guatemalan Indians, apparently misrepresented matters, perhaps lied, in her 1983 autobiography "I, Rigoberta Menchú: An Indian Woman in Guatamala" that advanced her cause. Apologists assert that "higher truths" are more important than the facts, that "there is no such thing as truth, only rhetoric".[25]

Political intrigues in other topics of scholarship may also lead to prestigious awards, despite questionable practices bordering on deliberate fraud. In 2000 historian Michael A. Bellesilles of Emory University won Columbia University's Bancroft Prize in History for a book advancing strange political notions that early settlers and militias of colonial North America were not well armed! Moreover, he posited that the myth of the armed frontiersman developed only after the Civil War!

Bellesilles directly became the highly lauded darling of anti-firearms advocates, but his scholarship is seriously flawed, perhaps even fraudulent, devised to support federal anti-firearms legislation. Many of his cited references are in error, do not support his contentions, or cannot be found at all! His original notes were "lost" in a flood. He has been abandoned by once overjoyed anti-firearms advocates and ordered to defend his work by his dean.[26]

Then there are other awards given for exceptional achievement. The Ig Nobel Prize Award sponsored by *Annals of Improbable Research, The Journal of Record for Inflated Research and Personalities* is given annually to those scientists who take themselves too seriously, for work that "cannot or should not be reproduced".[27] Yet other journals dealing with science humor, *Journal of Irreproducible Results* and *Worm Runner's Digest* are no longer published.

Among awardees of the Ig Nobel Prize in 1991 is Jacques Benveniste for his discovery that water has memory (see Chapter 5).[28] The 1998 winners include *inter alios* psycologist Mara Sidoli, Ig Nobel laureate in literature, for a paper dealing with flatulence "Farting as a defense against unspeakable dread".[29] Then there is the Sheldon Award named after Sheldon Hackney, the former University of Pennsylvania president (later chairman of the National Endowment for the Humanities) who failed to take action against militant students who stole or burned campus newspapers bearing thoughts that did not meet with their approval.[30]

7. Some Remedies

With these developments I sometimes feel like the poilu at Verdun, *"Il ne passeront pas!"*. How do we remedy present excesses; how do we retain balance in a strange environment; how do we keep cool in Hell. Clearly a satisfactory critical approach to the technical aspects of biomedical science must be retained, but also awareness of real world aspects of biomedical research must be learned by those seeking honest careers in experimental science.

One must identify a problem and devise some remedy. From antiquity thinking individuals recognizing problems have accepted the torment and punishment that independent thought brought in the earlier days. We may find solace in recalling the history of human intellectual development dealing with problems over the millennia. We are not alone in the modern world; we have the thoughts of Dante and the past to comfort us in the realization that human progress is made despite the adverse effects of the inmates of Hell *"c' hanno perduto il ben de l'intelletto"*.

The present deliberate use of deceptive speech is not new but must have existed from the invention of language, built into human nature. Confucius (K'ung Foo-tsze, died 478 B.C.E.) recognized the importance of "putting names right" (Analects 13:3): "On matters beyond his ken a gentleman speaks with caution. If names are not right, words are misused. When words are misused, affairs go wrong - - A gentleman is nowise careless of words".[31]

Then, Socrates (died 399 B.C.E.) is quoted by Plato: "for false words are not only evil in themselves, but they infect the soul with evil".[32] Further, recall the Bible: "And Ye shall know the truth and the truth shall make you free" (John, **8**,32). Later, at the emergence of the philosophy of inquiry in the face of dogma of the Roman Church we have the monks Pierre Abélard (1079-1142), Duns Scotus (*ca.* 1270-1308), and William of Ockham (1290/1300-1349), but also the founding by 1233 of the Roman Catholic Inquisition to ensure correct thinking.

Abélard admonished us to doubt, to inquire:[33] "By doubting we come to inquiry". "By inquiring we perceive the truth". He then gave us four rules that state matters concisely and that apply today, indeed throughout time: (1) Use systematic doubt and question everything; (2) Learn the difference between rational proof and persuasion; (3) Be precise in use of words, and expect precision from others; (4) Watch for error, even in Holy Scriptures.

Question everything! Be precise! Seek error! Note that these admonitions use the word "truth". We will not know when we may reach the truth in science. We do not even know that there is truth in science. We do know that by effort we can reduce error in our search for truth. As "the pursuit of happiness" in our personal lives, so we have "the pursuit for truth" as an article of faith in science.

William of Ockham gave us Ockham's Razor that reduces issues to their simplest terms: *"Frustra fit per plura quod potest fieri per paucioro"*. In vain are explanations with multiple hypotheses where single ones suffice. A more cited version: *"Entia non sunt multiplacanda praeter necessitatem"* appears not to have been used by Ockham.[34]

The modern systematic period was fashioned by Francis Bacon (1561-1626), an initiator of rational inquiry and of modern experimental science, who wanted the scientist to be a disinterested observer free of prejudgments and preconceptions. He remarked "Truth emerges more readily from error than from confusion".

Then there was Isaac Newton! As Alexander Pope noted "God said, Let Newton be! And all was light". Newton, who said "I can calculate the motions of heavenly bodies, but not the madness of people", codified his ideas into four rules for reasoning, of which the first two are cited here: *Regula I: Causas rerum naturalium non plures admitti debere, Quam quæ & veræe sint & earum phænomenis explicandis sufficiant.* We are to admit no more causes of natural things than such as are both true and sufficient to explain their appearance. *Regula II: Ideoque effectuum naturalium ejusdem generis eædem assignandæ sunt causæ, quatenus fieri potest.* Therefore to the same natural effects we must, as far as possible, assign the same causes.[35]

More recently, Bertrand Russell has given us his Liberal Decalogue of ten or so points as guidance in these matters, among which are: "Do not feel absolutely certain about anything!", "Never try to discourage thinking for you are sure to succeed", and "Do not fear to be eccentric in opinion, for every opinion now accepted was once eccentric".[36]

Times have changed in modern science and are changing yet. The insulting methods of politics and business in advancing individual special interests are now apparent in biomedical science as well, and students need to understand this development fully. Otherwise, as doubts arise, we may become victims described by Robert Frost:[37] "We dance round in a ring and suppose But the secret sits in the middle and knows".

One of the disadvantages of modern biomedical literature is the need to sift through trivial, erroneous, repetitive papers that add little or nothing of use or value but that must be examined for possible reserve use, or for remote interest. There just might be some jewel among trivia.

Many such papers aid the authors in their quest for fame, fortune, or just survival. Redundant publication of the same material in more than one journal, under different title or juggled authorship, is also a current problem, as some seek to gain fortune by this means. It is common practice to see the same material reported in abstracts at meetings, and in such circumstances we discover the famous Least Publishable Unit (LPU) commonly seen in biomedical literature. There may be more than one advantage to the perpetrator: (1) His bibliography is increased greatly by LPU; (2) On grounds that the Dean cannot read but can count, an expansive bibliography may lead to promotion, increase in amenities, space, and stipend; (3) Funding organizations are pleased to be honored so frequently; (4) Opportunities arise for inclusion of courtesy authors or exclusion of social offenders; and (5) Any individual LPU can be ignored after each is proven redundant, trivial, or just wrong.

It is ever so common to see encomia about important biomedical researchers in which a total number of published items is provided, numbers nearing a thousand in some examples. Clearly, the larger the number, the more important the person, the greater the contributions. By contrast, as heard on Golgotha: "I hear he was a great teacher". "Yes, but he never published anything!".

Such enterprises are aided by the increased number of science journals now published. Few can follow all items of potential interest. Journals of science societies are now in competition with those published by commercial presses. *Chemical Abstracts* presently abstracts items from over 8000 journals! Both society and business journals have massive increases in the number of pages published each year. The *Journal of Biological Chemistry* now publishes over 25,000 pages annually! Other journals in exponential growth publish many volumes per year, in one case estimated soon to be one volume every five minutes.[38] Undiscovered duplicate publication becomes an option to build one's bibliography.

Moreover, the information overload in the biomedical sciences prompted past NIH director Harold E. Varmus to propose creation of an electronic publication system PubMed Central (previously "E-biomed") that includes full texts of soon-to-be-published journal articles. Publications for high school and college undergraduates are now encouraged. It is as if a publication list is a requirement to be anybody in science. There is even a new *National Journal of Young Investigators* on the World Wide Webb of the Internet, supported by the journal *Science*, the NSF, and industry.[39]

With the vast expansion of publications to serve the author's needs rather than those of science, a direct result of "Publish or Perish", and the decline of adequate editorial review of simple English, grammar, and syntax in journal articles, coupled with an increase in errata now published regularly the practicing scientist has the burden of trying to keep aware of progress in narrower and narrower fields. Many a journal now carries sections titled "Addenda and Corrigenda", "Errata", "Corrections and Alterations".[40]

CHAPTER 4. BEWARE THE MEANING OF WORDS

War is peace
Freedom is slavery
Ignorance is strength
- George Orwell, 1984

As we careered through each set of assigned papers described in Chapter 2 many major points emerged, as each lesson had by design something to provoke thought. Some points were repeatedly noted; others had their impact in one lesson. On occasion class discussion diverged a bit from the assigned details in order to make points about items of universal application, as befit each circumstance.

Two major issues pervaded our thoughts as we reviewed the biomedical literature, issues receiving treatment in separate chapters here. The foremost is the meaning of words, a topic of concern from the ancient past. The second item is the appeal of miracle or revolutionary science, where we must simply ask can it be so. The two topics together, the misuse of words and terminology copied from business and political practices and of new discoveries that tempt disbelief, pose a formidable challenge to those encountering these matters in published biomedical literature.

So, we come to the first unscheduled lesson, BEWARE THE MEANING OF WORDS! This was my greatest emphasis, nay preaching, from the first day. Mine is not a new message but is one our students sorely needed. Problems of science with the meaning of words have existed from the early days of modern science, with Jöns Jacob Berzelius (1779-1848) remarking: "We make use of a word to which we can affix no idea". Later, Frederick Gowland Hopkins (1861-1947), Nobel Prize in Physiology or

Medicine 1929, noted *ca.* 1921: "They escape from the hard task of analysis by falling back upon the properties of an entity to which, since it is wholly imaginary, any attribute may be ascribed".

In ever so many places our students merely overlooked words in titles and text that they did not know, thus bringing down the wrath of God. When asked about "Colligative" and "Lyophile" in titles and "suckering" (botanical term) and "entatic" (excited ?) in texts, students were found ignorant in their first confrontation. Thereafter, students bothered to understand the meaning of words lest they be asked in class about meaning. Every word is important. How understand a paper not knowing the words used, not bothering to use the dictionary, with the loaded PC words that mislead. Lesson learned.

Here there was direct conflict with PC, PC designed to mislead and confuse and to assert authority over common sense. On occasion a student objected to the nonPC word "bitch" used appropriately for female dogs. Recall that by modern PC, if anyone feels offended by a remark or comment, then there is a real offense.

It was not our intention to offend but to enlighten; however, in trying to do so we had to say things that may seem extreme. Our remarks about individual papers were not attacks on the authors but were made to reveal error in its several forms. Any course in criticism must deal with several issues of today's real world, including the horror of PC that leads to error, to fear, to fraud, and to federal government intrusions and control.

There is a steady increase in science writing of terminology mimicking the deliberate political and business revisions of word meaning. Whereas our biology students were well versed in their own jargon and found no problems with newly minted terminology that will never appear in a dictionary, they had the expected troubles with vocabulary outside their specialties.

Some examples require little protection against misunderstanding; others are insidious, even dangerous if misunderstood. Once word misuse is recognized and understood, most are no longer harmful.

I shall review points made more than once in class as need or opportunity arose. Some may seem remote from graduate instruction and also unnecessarily cruel to PC, but this is how we did it.

1. Molecular Things

One of my favorite tirades on the meaning of words dwelled on the word "molecular". You may think you know what that means, but guess again. The word has been sequestered as a recognition word for cell biology and genetics studies. Our word. *Paròla Nostra.*

"Molecular" has become a buzz word denoting that the paper, journal, book, or person is a member of the high priesthood, a get-with-it modernist, usually with federal research funding. Such terminology is now found in the title of otherwise serious biomedical articles: "Lipoprotein Trafficking in Vascular Cells. Molecular Trojan Horses and Cellular Saboteurs".[1]

Among published uses of "molecular" are: "molecular chemists", "molecular anthropologists", and "molecular engineer", whatever they be. We see "chemistry at the molecular level" and "molecular" advances, approaches, biosciences, chaperons, computing, control, cycles, data, definitions, demolition, device, diagnosis, diversity, electronics, evidence, explanations, fossils, knot, logic, machines, magnets, materials, methods, movies, operating environment, organic solvents (!), remissions, research, rotor, shuttle, studies, switch, techniques, umbrella, wheel, and wire, *inter alia*! Mindless foolishness!

However, we pass from nonsense to error in the expression of molecular weight in kiloDaltons! In ever so many published biology papers the molecular mass of proteins is given as molecular weight expressed in kiloDaltons or kDa. Molecular mass is a measure expressed with units of mass (Daltons); molecular weight is a unitless ratio of masses. The difference in concepts appears to be unknown to biologists. I hammered on this issue at each need when discussing published work. Students had a hard time with this, as their biology mentors did not realize the difference and surely did not care. In one instance, one of our very capable junior biochemists even excused the error as "lab jargon".

My greatest glory came when one of our students corrected one of the biology faculty in a formal seminar. The biologician had mentioned a protein with a molecular weight of so many kiloDaltons, and the student pointed out this error. Right there in the seminar room! Not only was the message getting to the students, but one of them also had gained confidence in speaking out when error was rampant.

The error is not confined to biologists and has appeared in *Chemical & Engineering News*[2] and other chemistry journals. My letter to the editor on this matter was answered with the lame excuse that several "chemists" had reviewed the article and seen no problem.

Both science societies and commercial publishers find the term "molecular" a sure selling point. Be it molecular, it is good. "Molecular" books now issue regularly, Molecular Bacteriology, Molecular Diagnostics, Molecular Embryology, Methods in Molecular Medicine, and Principles of Molecular Medicine issuing from Humana Press in 1998-1999. I contrived a list of over 80 science journals with "molecular" in their titles, a select few being: *Journal of Molecular Biology, Journal of Molecular Catalysis, Journal of Molecular and Cellular Cardiology, Journal of Molecular Liquids, Journal of Molecular Medicine, Journal of Molecular Recognition*, and *Journal of Molecular Structure*, but also *Molecular Brain Research, Molecular Ecology, Molecular Engineering, Molecular Human Reproduction, Molecular Medicine, Molecular Microbiology, Molecular Physiology*, and *Molecular Physics*.

Which of these examples has a title that is appropriate

outside of club membership. No journal "Molecular Surgery" has yet appeared, as that might mean chemistry, the making and breaking of molecular bonds. In reality, "molecular" in most of these contexts merely means genetics nucleic acids/protein biology.

2. Current Practices

The following examples of current writing practices is incomplete, as there are ever so many other examples. The categories are arbitrary, and many items belong in more than one category. Among items in current public use, some of which now occur in science writing, are:

Simple Ignorance. Poor English composition through ignorance is not as bad as writing with deliberate uncertain, ambiguous, or confused meaning. One cannot do much about science papers published by ignorant persons save to note that where there is error from ignorance there is hope of learning, of improvement. Use of a good guide to English composition and a dictionary, taken together with careful study of well written research papers, can remedy matters.

Knowing the meaning of words is of prime importance. Misuse immediately displays the extent of the writer's erudition. Word confusions "alternate versus alternative" and "careen versus career" once standard elementary school material continue to be irritations. As for chaste versus celibate, recall the Florida politician who won election after proclaiming his opponent had practiced celibacy before marriage!

Error from ignorance, or more kindly from unawareness, abounds. Nonetheless, the only way to avoid error in the conduct and reporting of experimental science is to do nothing. Recall the Chinese axiom *"Wu Wei Ehe Wu Wei"* (Do nothing).

One must surpass error by correcting it. Even serious scholars err, witness an apparent error in rendering the motto of the honor society Phi Beta Kappa into Greek, where one letter and one diacritical mark may need revision for correct grammar.[3] Attention to such detail is no small matter, any more so than the careless misplacement of a decimal point, or "now" for "not", merely one letter different.

Carelessness. Carelessness seems to be a normal property of many students, particularly those with poor preparation for graduate work. Here again, there is hope, as students are forced to realize the immense importance of removal of error from carelessness in experimental and written work. Laziness and stubbornness must be considered here as well, as indifference to improved writing is encountered regularly.

Carelessness shows in many manuscripts submitted for publication; indeed, refereeing some manuscripts is so laborious because of carelessness that unpaid referees may simply stop and not complete the job. Many editors do not correct careless writing, and the matter is there for all to see, if there is any interest. Upon reading a carelessly written article, how are we to regard clear evidence of carelessness? Were the experiments conducted with the same carelessness? Should we bother to read further?

Among the many careless writing errors is the heavy use of laboratory jargon, e.g. "hot" for radioactive labeled material and "cold" for unlabeled compounds. In-house terminology, words with meaning to the users but with none to the rest of the world, code words used for special effect and for security purposes, foreign language words, undefined acronyms, and obsolete terminology have no place in published works.

Code words carelessly used may disclose information, even company secrets, in unintended fashion. The extent of research progress may be revealed by the proprietary code numbers assigned research compounds under testing. In 1956 triamcinolone was assigned the code number CL-19824 at Lederle Laboratories, in fact revealing the extent of the company's testing at the time. In contrast, the Wyeth Laboratories drug norgestrol coded WY-3707 disclosed the very limited extent of the company's research at the time 1963. In sharp contrast the latest disclosures by Merck & Co.

of a microbial agent potentially effective against diabetes describes the material as L-783,281, thus indicating their very large number of compounds of research interest.[4]

Fashionable Foolishness. It is currently fashionable to see and hear all sorts of fad cliches, nonce words, and slang in our lives. However, these same activities now regularly found in the biomedical literature may lead to confusion of intended meaning. It appears some authors and speakers have so little erudition that reliance on such terminology simply follows from their low state of scholarship. One major fault lies in the use of undefined abbreviations, acronyms, and "cute" items now the fashion in biomedical publications.

There are many standard acronyms for important cell constituents that no longer require definition at first use in most journals: ATP, NAD, NADP, DNA, RNA, and the like. However, the use of acronyms in published "molecular" papers surpasses common sense. If the whole point of publishing work is to provide the reader with information that can be perceived by reading, the use of acronyms, defined or no, causes no end of difficulties.

Until very recently only the cognoscente or molecular biology priesthood could read some papers, so many are the acronyms. Our favorite uses 17 acronyms or abbreviations on the first page, some undefined.[5] So great is the problem that there is now a published guide to these matters, thus ensuring those who use the guide some degree of success in communication of their work.[6]

The acronym problem is not confined to biomedical articles, as it is now fashionable to devise cute acronyms for NMR spectroscopy pulse sequences. One simply must know from other information what COSY, NOESY, SECSY, HOHAHA, and AMNESIA, *inter alia*, mean. Acronyms that spell unsuitable words in foreign languages should be avoided; recall AMGOT (American Military Government of Occupied Territory) from World War II, shortly changed to AMG.

As for the cliche, examples regularly seen and heard in public discourses are "point in time", "real time", "laced with cyanide", "pre-" words ("prereserved", "preowned"), "recorded live" (recorded before an audience), and "infrastructure", *inter alia*. Many others in popular usage are best avoided completely; all such terms do is detract from the message of the authors.

To emphasize such lapses I asked the students the meaning of older nonce words "23-skidoo", "goo-goo eyes", "zoot suit", and "feather merchant", all totally unknown terms now. Obsolete science terminology may also become unreadable after a while; recall γ for μg, λ for μL, mg% for mg/100 mL, terms once broadly used. On this matter it also came out that students do not recognize the terms "1-A" or "4-F" now that the military draft is no longer.

Clearly slang has no place in serious science communication, but jargon now regularly emerges in the biomedical literature as a suitable means of expression. Also, it is now a popular fad to use "cute" terminology, words that presumably disclose humor or smartness, that establish your membership in the club. We see the terms "knockout mice", "nuclear run-on", "footprint", "library" (collection), "lariat" (a particular cyclic structure of the "spliciosome"), "scafold" and "platform" (basis, framework), "CAT box" (a specific DNA region), "leucine zipper" (a leucine-rich region of protein), and "zinc finger" (a DNA binding domain in protein containing zinc bonded to two cysteine and two histidine residues). Zipper and finger had earlier uses: the German MG42 machine gun of World War II was "Hitler's Zipper"; there were "polar zippers" of macromolecules associated with Huntington's chorea. The "proactinium finger" was a region of mass stability for superheavy nuclides.

Other recent cute terminology involves "orphan" receptors, nuclear proteins that may bind some unknown agent (hormone, ligand) to become ligand-activated transcription factors that then bind to the regulatory region of a target gene to affect gene activity. As the activating ligand and associated gene activity remain unknown, receptor status of these proteins is inferred from their amino acids sequence suggesting a DNA-binding domain. Accordingly, these proteins are thus putative receptors, but "orphan" receptors is so much more the cute term.

The same levity appears in chemistry. Recall the "barn", the nuclear neutron capture cross section of 10^{-24} sq. cm. dimension, a probability of nuclear reaction. "As big as a barn". So it was. We also have "color", "charm", and "flavor" properties of subnuclear particles. In a recent convergent total synthesis of fungal tremorigen penitrem D two intermediates to be joined are termed "western and eastern hemispheres".[7]

Yet another cute development lies in the use of assertive complete sentences for the titles of "molecular" journal articles. It is now fashionable in respected biomedical science journals to tell readers in the first thing they read, the title, what is so. One no longer even needs to read the summary or abstract of the paper, let alone the text, to know what is true. The assertive sentence title robs one of exercise of one's own judgement, so far has modern science progressed.

James D. Watson used assertive sentence titles for chapters in his 1965 textbook The Molecular Biology of the Gene, and his stature may have influenced once more the pathway his science took. The assertive sentence title is offensive in its nature, and the growth of its use is disturbingly a sign of the necessity of attaining a preset goal for the research reported. No longer do we see titles "On the Thus and So"; we must have positive conclusions lest credit and priority be lost and future research grant applications be unfunded.

On another point, the repetition of trivial or erroneous assertions in journal articles now occurs regularly. How many times must the same trivial or incorrect information be included in a published science paper. Any redundancy is excessive, but repetitious assertions reveal at least three features: some authors must repeat themselves in order to believe their own remarks; some authors deliberately attempt to sell a point that is not so; there is a lack of serious editorial attention to the paper.

Recall the membrane sugar transport protein composition values and estimated Polywater bond energy values previously mentioned (p.37, p.62) as examples of such foolish enterprise. Did anyone review the finished manuscript before submission? Did any editor look at the text?

I held forth a simple proviso to our students who fell into the clutch of foolish use of words. English is a rich language, and there is always a simple, well understood English word for the nonce words of the moment. One does not need to alter English any more than sensibly required for clarity of meaning. Although we can read Chaucer today with some understanding, that is not so for the earlier Old English text of Beowulf, written down from oral traditions about the year 1000 C.E.

However, there is always a day when a new word is needed for new developments (radio, television). The recent appearance of the new word chaperonin for those protein complexes that regulate folding patterns of other proteins is a modest exercise of molecular biology creativity. One need not go to excesses in inventing new words or creating pretentious neologisms, but new words will be had, extreme or no.

Humor may intrude in some inventions, as in Buckminsterfullerene (a.k.a. buckyball, footballene, now just fullerene) for the recently discovered third C_{60}-form of elemental carbon and in "Swisscrossane" for cyclododecane. Humor is also expressed in other ways, including a diagram of laboratory apparatus in which there is a tiny fisherman.[8] Another be in acknowledgements, the five year old daughter of E. J. Eisenbraun being thanked for her donation of *Pogonomyrmex californicus* ants from her Christmas ant farm gift.[9]

Further, no new long Greek words are needed. In 1954 the word synthetolikmisis (meaning compound winnowing) was proposed for the growing use of the word chromatography in the literature.[10] Neither should we use words with other meanings in foreign languages. Methadon, now spelled methadone, was called "amidon" (French for starch) or amidone in early papers.

As opposite to serendipity, the happy accidental discovery of something nice, there is now "bahramdipity" devised after King Bahram Gur of Persia (*ca.* 418-428), meaning the suppression of a discovery, perhaps a serendipitous one.[11]

Another problem lies in the misuse of terminology long established in one branch of science but twisted to other use. Thus, "halflife" well established as an exponential kinetics term describing rate of chemical reaction now means in "molecular" speech that time it takes for the life or lifetime of a species to decline by half, with the life of the species being twice the "halflife"; quite a different meaning! Those unaware of the terminology differences of the "molecular" versus chemical sciences may remain perplexed.

This sort of confusion is not confined to biomedical science. Another example of confused use of the same term for different meanings lies in "cold fusion", another binary word combination potentially misleading as to meaning. Cold fusion as initially defined involved the nuclear synthesis of superheavy transactinide elements by processes developed at Dubna, USSR, by Yuri Oganessian beginning *ca.* 1975.

Rather than using neutron or light element projectiles, heavier elements from argon to germanium are used, fusing upon collision with yet heavier target elements to yield new superheavy nuclides. Genuine nuclear fusion at low temperature is thus a reality, but one requiring expensive nuclide accelerators.[12] However, the term is now more popularly applied to the sensational alleged release of excess energy in the form of heat from inexpensive electrochemical cells having a palladium metal lattice as cathode bathed with deuterium ($^2H_2$) gas, a process regarded as unreproducible if not unreal (see Chapter 5).

Euphemisms. The euphemism intended to diminish the potent impact of certain words has its place, and in polite society kindness in speech is well received. One avoids skatological terms. Besides the long-time substitution of bathroom for toilet, or water closet, we hear "doo-doo" used publicly by a president of the United States. We now also see in advertisements "dried plums" for prunes, thus avoiding snickers about bowel action. More ominously we hear of "greeters" for muggers in New York City.

Other current euphemisms frequently used are: preowned (used), pass away (die), minorities (Negroes), sex (sexual intercourse), to have sex (coitus, to copulate), sleep with (sexual intercourse, copulate with), oral sex (cunnilingus, cunnilinctus), anal sex (sodomy), and gay (homosexual). With a good English word for specific meaning, why resort to such euphemisms.

Euphemisms abound in Orwell's 1984 Newspeak, in modern PC, and in Spookspeak, the language of our Central Intelligence Agency. Newspeak gives us the prize euphemism "joycamp" for forced labor camp! Current PC needs are accommodated in "confidence course" for the old obstacle course of military training, revisions of the Bible with "God the parent" and Jews as "religious authorities". Spookspeak includes "terminate with extreme prejudice" (assassinate).

The euphemism is less a problem in biomedical science writings, but we do see "sacrifice" frequently in ever so many papers. Whenever experimental animals are to be killed, they must be sacrificed. This actually may be required by granting agencies now, as all too many other requirements are imposed on experimentation with living vertebrates. Worms and plants are excepted from protection.

Also a form of euphemism is the overt sanitization of literature, art, and song imposed by church and state long with us, now seen as an aspect of media-promoted PC as well. The use of a word that hurts someone's feeling is now an offense, and a sanitization process must be in place to avoid such matters. Sanitization denies us the original words of serious authors who had no harsh motive in using their words in meaningful fashion of their times.

The word "nigger" in Mark Twain's serious fiction Tom Sawyer and Huckleberry Finn draws uncommon vituperation for the designed use of regional speech, but what can be done with Joseph Conrad's novel The Nigger of the Narcissus. Luckily, this story is not known to the sanitizers. The Gilbert and Sullivan operas with the offensive word must be sanitized; recall The Mikado ("nigger serenader" now "banjo serenader"; "stained like a nigger" now "painted with vigor"), Princess Ida ("And the niggers they'll be bleaching"), but also Trial By Jury ("Be firm, my pecker"). We know the song "Old Man River" has been cleansed of the word nigger, but "old" and "man" are now under revision by PC aficionados. We must say "senior citizen" for "old" and "person" for sexist "man", thus making the song Old Person River or Senior Citizen River, and perhaps eventually Paleoperson River!

Cole Porter's song "You're the Tops" at first had reference to Benito Mussolini, "You're Mussolini" as a positive matter since altered. The Studebaker automobile "The Dictator" model could not be sold today. Even our glorious national anthem must be revised, the third verse uncomplimentary to the British and Hessians removed and the powerful fourth verse altered to "Oh, Thus be it ever, when free PERSONS shall stand". The original word "men" is sexist.

Sheer Idiocy. Sheer idiocy has yet to pervade biomedical science literature, but as with other influences, what is in the public domain may shortly be seen in more serious works. Three unbelievable idiocies are current: "I could CARE less!", teacher training but driver education, and Ebonics. The unfathomable idiocy of uneducated football announcers exclaiming "I could CARE less!", meaning "I could NOT care less!" requires accentuation of CARE for the effect. Were the ejaculation stressed on the word could, thus "I COULD care less" we would distinguish directly the real meaning. How this idiocy has endured is a lesson in itself, as it also appears in print media where the required accent cannot be indicated.

Incomprehensible Binary Word Combinations. The combination of two words creating new special meaning incomprehensible from the words alone is a popular method of controlling thought processes. The assignment of new meaning is not confined to biomedical science but may be a universal remedy for containment of meaning among the inner circles. What would a first encounter with "quantum mechanics" mean to chemists and physicists. Or, atom bomb! Of course, our prize example of this matter is "molecular biology" itself, a term coined 1938 by mathematician Warren Weaver. So well entrenched is the term now that its meaning is no longer obscure, although some definitions leave us wanting.

In 1953 the discovery of the dimeric nature of DNA by James D. Watson and Francis Crick revived the term "molecular biology", perhaps an unsuitable term but surely a catchy name that has persisted and flourished. Were the two words hyphenated or just one word as in German, the new terminology, molecular-biology or molecularbiology, would make sense, but by keeping two words a whole outgrowth of molecular matters continues to expand. My message is probably lost on true believers and biology students who cannot understand why words are so important when we all know what they mean.

Molecular biology, the study of living systems at the molecular level, has become a new religion among its practitioners. The beatitude "Blessed are the molecular biologists, for they shall be given grants" is a reality, so much so that the basic underpinning must be ignored in preparation for entry into the priesthood. Yet others decry the matter as practicing biochemistry without a license. In mild annoyance of this status I had a sign put on my office door announcing the name of the laboratory "Radical, Ionic, Molecular, Atomic, and Chemical Biochemistry", that appellation appearing to exhaust possibilities (we did no nuclear or subatomic biochemistry, whatever that might be).

Other two-word terminology currently in the public vogue includes: affirmative action (race/sex preferences and quotas), sexual assault (previously rape), dynamic entry (unannounced forced entry by police), and sexual harassment. No certain meaning can be attached to such terms as sexual harassment or sexual assault, as any meaning is flexible, depending on nonce whim and fancy. Even the well understood word rape is redefined by the American feminist movement to include all sexual intercourse: "In a patriarchal society all heterosexual intercourse is rape"; "rape exists any time sexual intercourse occurs when it has not been initiated by the woman".[13] Then there is "visual rape", the ogling of a female by a male.

Switching Words/Meanings. Insidiously switching terminology for unearned political advantage is well known. In the recent case in Houston,TX, a local referendum election was overturned June 1998 because of switched terminology on the ballot. The approved initial version stated "The City of Houston shall not discriminate against or grant preferential treatment to, any individual or group on the basis of race, sex, ethnicity, or national origin". However, the printed ballot asked instead "Shall the Charter of the City of Houston be amended to end the use of affirmative action for women and minorities?", thus a perfect example of PC manipulation. National news media that carried the election results defeating the altered proposition did not report on the cancellation of the election by a federal judge because of switched wording.[14] The City of Houston has promised to appeal; a new election will cost more than $1 million.

We note that switched terminology is often involved in what should be serious discussions between scientists and religionists. Three words regularly suffer: faith becomes religion, reason becomes science, evolution becomes origin of life. Adherents of "creation science" and "intelligent design" regularly enter the fray with the word evolution but insidiously switch their focus to origin of life, thus revealing their true problem with science. One may think that a discussion involves evolution, but the switch to the origin of life is made directly, making further meaningful discussion of evolution impossible.

Deceptive switching of words also occurs away from political and religious matters. However, we must forgive the deception of code words necessary to wartime national security. The British of World War I coded their armored, tracked, armed vehicles as "tanks" to maintain secrecy, and the term stuck. In the days of our wartime Manhattan District nuclear weapons development "barium" was used as code word for uranium, "copper" for plutonium. Where barium or copper was really the element involved, "Honest-to-God Barium" or "Honest-to-God Copper" was used.

Another example of deceptive switching bears recounting here. The work of J. G. Bednorz and K. A. Müller that demonstrated superconductivity of substances at temperatures above that of liquid helium, for which they were awarded the Nobel Prize in Physics, led to hyperactive, competitive searches for yet other substances superconducting at higher temperatures.

By story, Ching-Wu ("Paul") Chu, University of Houston, submitted an account of his work with new superconducting YBaCuO substances for rapid publication but had YbBaCuO in the typescript, switched from element symbol Yb (ytterbium) to Y (yttrium) in the proof, ostensibly to mislead anyone using his information inappropriately before publication. Again, by story, several masters of exploitation using the privileged information could not reproduce results using ytterbium. By these switches and codes the YbBaCuO preparations should be Yttrium-Honest-To-God Barium-Honest-To-God Copper-Oxygen! The University of Houston increased Chu's personal salary and funded his development work handsomely.[15,16]

Moreover, there is the real matter of serious misunderstanding of the meaning of words, sans duplicity. The problem is not new in biomedical matters as we have a 1953 article "Science at Bay" in which Alice in the Wonderland of research encounters a director of research recruiting a chemist: "the real need of our company is for basic scientific research". Some time later, as results fail to please: "Yes, but not that academic type of research", "We do want basic research, but not that type". Finally, "We believe in the group research approach", "We will select a well known clinician... and let him hire as many chemists... as he needs".[17]

The more serious aspects of switching terminology in mid-presentation mentioned earlier may be accompanied by science humor in other switches of note. The ever increasing sensitivity in detection and measurement of trace chemicals has given us two new SI prefixes for dimensions: zepto for 10^{-21} and yocto for 10^{-24} (also Zetta for 10^{21}; Yotta for 10^{24}). However, a proposed switch from yocto- to guaca- has been advanced for 10^{-24}. Now, recall that a mole of substance contains the Avogadro number 6.023×10^{23} of atoms or molecules. Thus, by the proposed switch from yocto- to guaca- we have the guacamole, 10^{-24} of a mole of substance, containing but 0.6 atom or molecule, by the *Avocado* number!

The guacamole terminology was advanced as early as 1975 as a possible aid for attendees of the Chemical Congress of the North American Continent in Mexico City. The terminology though amusing has a serious side, as it is now possible to detect the optical absorption spectrum of one molecule. W. E. Moerner used the guacamole in lectures of one molecule chemistry from 1991, tried to use the term in a *Science* review in 1993, and finally got the term published in 1996.[18,19]

Political Strife. Although overt political strife has yet to be important in the biomedical science literature, the imposition of political notions on the conduct of biomedical science occurs from time to time. Thus, the American Society of Biochemists and Molecular Biologists cancelled plans for the 1993 annual meeting in Colorado because of a state referendum, passed by the people, that failed to provide legal favor for homosexuals. Failure to politicize may be as great a problem as politicization!

In the public domain there are ever so many terms currently used regularly for unearned political advantage, in both benign circumstances but also in aggressive matters. Among such items are: "discrimination" (unpleasantness to females and to Negroes), "diversity" (special rights for favored races and classes), "assault weapon" (self-loading military style firearm), "assault rifle" (select fire machine gun, but now also an assault weapon), "Saturday Night Special" and "junk guns" (inexpensive guns subject to prohibition), "right to life" (anti-abortion), "freedom of choice" (pro-abortion), "hate crime" (special protection for favored class victims of crime), "Antisemite" (anti-Jew; other Semitic peoples not included), "freedom of speech" (freedom of actions, flag burning, supporting political parties with money), "cold pasteurization" (irradiation of foods), and "Peoples Democratic Republic" (communist state). Use of these and related terms appear not to have invaded serious writings of science that are not devoted to political strife.

Present partisan warfare between our two major national political parties gives us: "greed" meaning low taxes, "compassion" meaning high taxes; "fairness" meaning state-enforced equality, "unfairness" to mean an individual right to self betterment.

Government Influences. Our federal government influences biomedical science through control of research funding, but the occasional politicization of NIH, Food and Drug Administration (FDA), and other agencies tends to bring official sanctions into focus. National governments have exerted influence in many ways, including proffered services and subsidies but also government sanctioned language, revision of history, propaganda, and ultimately despotic force. The use of subsidies to control events is age-old. The bribe paid to adversaries, the hiring of mercenaries, and issuance of contracts and grants are well known functions of government to affect its purposes. Control of language is of vital importance, witness the semanticide techniques of the German Third Reich and of the Soviet

Union in redefining words for propaganda purposes[20] and the total replacement of standard English (Oldspeak) with Newspeak in Orwell's 1984, *vide infra*.

In revising history we note Lev Trotsky, creator of the Russian Red Army in the Civil War period, became a nonentity and disappeared from history following his loss to Josef Stalin in 1927. Stalin had Trotsky murdered in 1940, thereby to ensure no arguments about who was the true Soviet hero. Even today modern Russian history books providing names and details of limited participation of minor Revolution and Civil War persons ignore the Trotsky name and deeds.

Following the Litvinov-von Ribbentrop treaty of 1939 between the USSR and Germany both governments moderated their political actions against the other somewhat. Previously, the words fascist and fascism were terms of opprobrium in the USSR for Germans of the German Third Reich and for Italians under the Mussolini government, but the word fascist disappeared in the USSR 1939-1941, only to return upon the German attack on the USSR. In turn, the word Stalingrad disappeared in Germany after January 1942. Stalingrad, once Tsaritsyn, is now named Volgagrad.

History has always been written by the victors, and we see this principle in action today wherever PC or its international equivalent abounds. The revision of American history that emphasizes that Columbus be a harsh exploiter of American Indians (oops, Native Americans) preoccupies some PC advocates beyond good sense. Columbus was not different from other conquerors of past recorded history, but for political hay Columbus has been selected for revisionist history.

More damning is the refusal of Japan to acknowledge the facts of their military government conduct of World War II. Revisionists siding with the Japanese position have now proposed that the pejorative term "Bataan Death March" of surrendered American military personnel on Luzon in 1941-1942 be appropriately replaced with the "Bataan High-Mortality Promenade" for the American guests accompanied by Japanese soldiers who called "bonsai" (not "Banzai") offering dwarf trees to the Americans.

The use of naked power by government and religion to quell defiance and heresy is also an ancient activity. We know of Socrates condemned to death by drinking hemlock 399 B.C.E.: "Socrates is an evil-doer, and a curious person, who searches into things under the earth and in heaven, and he makes the worse appear the better cause; and he teaches the aforesaid doctrines to others".[21]

Among other signal achievements of religion is the case of the monk Roger Bacon (*ca.* 1214-1294) imprisoned by his own Franciscan order for fifteen years 1277-1292 for his remark of 1267 "The calendar is intolerable to all wisdom" suggesting the Julian calendar then used by the Roman Church was inaccurate.

Anent treatment of scholars and scientists by totalitarian regimes, we recall the legend of Galileo Galilei, when shown the instruments of torture by the Roman Church, retracted his heresy in order to avoid consequences. Yet we are told he murmured *"Eppur si muove "* ("And yet it moves").

More recently, genetics research suffered greatly in the USSR. We recall the death of accomplished Russian geneticist Nikolai Ivanovich Vavilov who perished 1943 in the GULAG joycamp where he was imprisoned for failing to accept Trofim D. Lysenko's version of Soviet genetics. Lysenko promoted vernalization of wheat and Lamarckian inheritance of acquired characteristics to the state's satisfaction.[22] In contrast to Galileo, Vavilov, since rehabilitated posthumously, is quoted: "We'll go into the fire and burn, but we won't renounce our convictions". The less known geneticist Dmitry K. Belyaev merely lost his official appointment as head of an orthodox genetics animal breeding program in 1948.[23]

Forget not Elena Ceauşescu, a self-proclaimed famous biochemist who dominated Rumanian biochemistry and organic chemistry from her exalted political status and who was executed with her husband Nicolae Ceauşescu,

Communist Party dictator of Rumania.[24] As his wife she was accorded great privilege and many honors during her reign. She regarded herself as a prestigious scientist and took every opportunity to emphasize her importance. She appointed herself to the Roumanian Academy of Sciences, sought foreign recognition and honors, but was a bully and harsh, mean-spirited director of the Institutul de Cercetări Chimici which she headed from 1965. She put her name as first author on every paper. No conference could be held without her name being prominently first. The whole 1989 year issues of the *Revue Roumaine de Chimie* were dedicated to her: "In Honor of Elena Ceauşescu, D. Chem. Eng. Member of the Academy of The Socialist Republic of Romania".

Yet Elena Ceauşescu was a cruel fraud. She became a chemical engineer after two years at Muncitoreaşca University where uneducated Communist Party members obtained degrees. She was graduated only after political pressure was exerted, after two prior failures. She then sought to study organic chemistry but was refused. The professor lost his grant thereafter and access to the science literature. Next she worked in the Roumanian Academy of Sciences, where she defended her Ph.D. work *in camera* in 1970. By law and university tradition doctoral defenses were open to the public; hers was closed. Her professor was thereafter demoted as director of the organic chemistry center he had founded.

There are words in her published papers she could not pronounce, let alone understand. She was unable to recognize the molecular formula H_2SO_4 for sulfuric acid, did not know what a chromatograph was, and in other ways disclosed her ignorance of science.

Newspeak. In George Orwell's novel 1984 we learn of Newspeak, the epitome of government control of people, the means of attaining and securing English Socialism (Ingsoc).[25] Newspeak was designed to replace Oldspeak (Standard English) by destroying the meaning of words, by making it impossible for want of available words to think or speak unacceptable thoughts. More important to our consideration of biomedical science, in Newspeak there is no word at all for science! The goodword Ingsoc says it all. Rational inquiry of Nature thus became impossible.

Such Newspeak slogans as "War is peace, Freedom is slavery, Ignorance is strength" appeared, leading to such pronouncements as "Four out of every three persons is an enemy of the people". Neither the brutal *"Arbeit Macht Frei"* slogan at the gate of the Dachau Lager (a revision of the earlier aphorism *"Wissen Macht Frei"*) nor Dante's motto at the gate to Hell *"Laschiate ogne speranza voi ch' entrate"* approaches the cynicism of Newspeak. Newspeak and the design behind it are genuine horrors.

'Twould appear that a similar plan to restrict unsocial thoughts and speech in the United States is afoot. Publishers of English dictionaries are one by one eliminating words deemed socially unacceptable, practicing logocide. The words dago, kike, wop, and wog disappeared from Webster's New World Dictionary, Second College Edition in 1970. Others excluded must surely include canuk, pollack, kraut, frog, yid, hebe, and spic. Gringo, honky, and cracker are not excluded, as these words refer to those of unprotected Anglo-European ancestry.

Current domestic racial warfare demands deletion of the N-word (nigger) on the basis "If the word is not there, you can't use it"! Shades of Newspeak! There may result a lawsuit if someone has their feelings hurt after seeing the N-word in a dictionary.[26] Similar removal of the F-word (fuck) but not of words of the same ancient meaning, such as swive, must occur as well. Such older words are not that well known inside or outside of the dictionary in any event.

There is historic precedent where words were proscribed. Ancient Jews were prohibited from using the name of God; ancient Germans had no word for "bear" so great was their fear of bears. More recent taboos abound. Mere mention of the word "leg" was enough for censure in Victorian England; limb was the appropriate word. By contrast, today the polite Victorian words intercourse, erection, and ejaculation may be misunderstood.

Although Newspeak has not survived as a modern language, its effects survive in the designed misuse of words rampant via the agency of PC, next discussed, that shares many characteristics of Newspeak. The same intended results are had, the deception of people and increased government control of everything, including science. Forget not Clintonspeak.

Political Correctness (PC). Political Correctness now pervades public life in the United States. Although PC has not been established as the official language of modern biomedical science, there are little inroads and hints of what is to come should PC prevail. The current PC emphasis on race and sex coupled with the growing use of DNA analysis for identification of persons, for definition of racial origins, and for designing treatment of medical disorders makes it inevitable that PC will influence events if unchecked by common sense.

It is essential to define PC. Perhaps people think they know what PC means. Some will surely subscribe to the PC concept in keeping with their personal educational level and political leaning; others may be less sanguine, may even insist on changing the acronym to TC (Thought Control).

Political Correctness is understood to be the revision of standard English language imposed by government, print and broadcast entertainment and news media, and political groups to meet their current nonce goals or agendas, employing appealing terminology with hidden agenda meanings advocated by these pressure groups. Political Correctness is the deliberate misuse of words to mislead, to confuse issues, and to advance personal or political advantage through deception. Methods used are emotional appeal, media support, intimidation, coercion, and government sanctions.

Comparison of PC with Newspeak is inevitable. Both tongues are insidious in their effects on civilization. Both remind us of Humpty-Dumpty's remarks on the other side of the Looking Glass: Humpty-Dumpty: "When I use a word it means just what I choose it to mean - neither more nor less". Alice: "The question is whether you can make words mean so many different things". Humpty-Dumpty: "The question is which is to be master - that's all".[27]

NEWSPEAK	PC IN AMERICA
Newspeak = official language	PC = official language
Oldspeak = Standard English	Sexual assault = rape
Ingsoc = English Socialism	Sexual harassment = Who knows ?
Goodthink = Orthodox thought	Gender = sex
Crimethink = Thought crime	Sex = sexual intercourse
Oldthink = Unacceptable thought	Gay = homosexual
Thinkpol = Thought Police	Facilitator = teacher
Goodsex = chastity	Normal = invidious term, do not use!
Joycamp = forced labor camp	Spin = lie
Uncold = warm	Linear = unimaginative, Old fashioned
Prolefeed = entertainment, news	Nonlinear = perceptive, avant-garde

But, there are differences. Newspeak has no word for science; thus, science cannot exist. In PC we have all sorts of science: political science, "creation science", and government regulated science. However, PC in the science literature obscures meaning, meaning not always understood from context. Also, there is no PC word list, dictionary, or thesaurus, thus posing greater damage to clear meaning than the cute acronyms of molecular biology.

These deceptions inevitably lead to misunderstanding, misunderstanding to error, error to fear of consequences, and fear to misconduct and fraud, all issues now of crucial importance to experimental science. In science, error may come in so many other ways. If we are already unsure of what we see in Nature, if our best science remains provisional until confirmed and woven into our best understanding, how shall we regard deliberate misuse of defined words to mean something else.

Anent fear, besides fear of loss of a fat federal research grant, there is fear engendered by PC and legal concerns. The innocent misuse of words not sanctioned by government, pressure groups, and indoctrination media may generate a problem. One is reluctant to use precise but banned words in lectures, journal articles, and federal research grant applications, lest charges of insensitivity be made, charges leading to penalties such as dreaded sensitivity training!

Two topics are particularly sensitive, race and sex. Those who misuse the three word farrago "Negro", "sex", and "gay" create a serious PC offense. Negro is banned in favor of "African-American". The language of President Harry S. Truman and Rev. Dr. Martin Luther King Jr. is no longer welcome or tolerated in PC America. In my medical lectures on drug metabolism I used an older slide showing differential responses of diverse human racial groups to inactivation of isoniazid, a drug of choice in treating tuberculosis. Something a physician-to-be might want to know, that race may be important in patient management. However, the slide made well before PC appeared used the Oldspeak word "Negro" and not the currently required PC "African-American" now prevalent in works seeking to maintain race consciousness. Although tuberculosis is a world-wide medical problem and not all Negroes in the world are "African-American", I was admonished not to use the slide. I wondered when an activist student protest might arise. I had no liability insurance for attack litigation as do some other teaching faculty. At my retirement drug metabolism was dropped from our medical lectures.

Moreover, it is no defense to be punctilious in the use of language, as those hearing or reading your words may misunderstand and provoke matters. In January 1999 the use of the word "niggardly" by David Howard of the Washington City,DC, government was taken by Negro city government members as insult, forcing Howard's resignation.[28] The word "niggardly" has nothing to do with race, but to those ignorant militants with low levels of comprehension this was an overt insult nonetheless. One now creates an offence merely by using words that sound like proscribed words.

Racial PC issues are supported by the federal government in other ways as well, some creating a class of thought crimes. The Office of Civil Rights in the Department of Education has issued harassment guidelines whereby schools must take immediate action if a teacher, student, or other person makes a racially offensive verbal statement. Dreaded sensitivity training is among punishment for infractions. The Office of Civil Rights does not restrict free speech, only "verbal conduct"![29]

Furthermore, the federal government deems it necessary to call American Indians "Native Americans"; indeed, the government has pronounced that Columbus named the native peoples he found in America "Native Americans". Little it matters that Columbus named these people "Indians" in his mistaken belief that he had found the East Indies of Asia. Does anyone remember Amerigo Vespucci who landed on the South American coast, or that his travels prompted Alsatian professor Martin Waldseemüller in 1507 to suggest the name America after Amerigo Vespucci. Facts should never compromise PC.

Now we see American Indian activists objecting to the names of some football teams as racial harassment. Binney & Smith Inc. makers of Crayola will rename their Indian red crayon chestnut despite the fact that Indian red in this case comes from India. Mere deletion of the terminal letter "n", making the name India red, apparently was too simple. Moreover, the song "I'm an Indian, Too" was excised from a recent revival of Irving Berlin's musicale "Annie Get Your Assault Weapon". Harassment no longer involves serious issues but now extends to trivial and symbolic matters.

Besides the race issues the words gay, sex, and gender typify problems of meaning mandated by current PC practice. Gay no longer relates to gaiety but is euphemism for homosexuality. Sex now means coitus, sexual intercourse, not the biological male and female sex. Gender now means sex in PC English and no longer is a grammar term that includes three classes masculine, feminine, and neuter. However, PC ignores the neuter, thus discriminates against homosexuals, transsexuals, and those excluded from the masculine and feminine groups. 'Tis wicked to use such divisive words.

Science journal editorial policy about sex is now infected with PC, as a journal editor asked me to change the word "sex" to "gender" in an expression describing experimental animals: "diet, age, sex, disease, obesity". My published article retains the normal meaning of sex.[30] We realize that the word "normal" now causes intestinal borborygmy among PC devotees.

Publication of other articles misusing the word "sex" may lead to much more severe consequences and to fear for one's appointment or job. In January 1999 the American Medical Association summarily dismissed George Lundberg, long-time editor of *JAMA* (*Journal of the American Medical Association*), for publishing an article "Would You Say You Had Sex If..." at the very time that President William Jefferson Clinton was impeached and under trial before the U. S. Senate on matters that included sexual activity. The article dealt with a 1991 survey among college students as to their understanding of the terms "oral sex" and "having sex". Whether inadvertent or cunningly planned, the article offended political supporters of President Clinton, thus the need for abrupt action. The journal's action gives new meaning to the term "political science".[31]

Another phase of sex foolishness is the removal of the term "man" from such words as chairman, fireman, or policeman, etc. Although the words human and woman remain unaltered, huperson and woperson being a bit much, a chairman has become a "chair"! The adjustment also applies to the plural "men", to personal pronouns he, his, and him, and the like, all to be avoided under PC. As substitute several new words have been invented: he or she becomes "she", him or her "herm", his or hers "hs" (pronounced zzz), man or woman "maman", boy or girl "birl".[32]

A related effort by me to adjust matters during the Equal Rights Amendment (ERA) crusade in 1977 involved submission of a personuscript "ERA Nopersonsclature" outlining what must be done in science to meet ERA requirements. The personuscript was accepted by the publisher George H. Scherr January 26,1977, but has not been published, thus justifying my claim for the longest wait for publication of a serious personuscript of record![33]

Although we have not seen it yet, demands for changing the heinous names of the elements helium to itlium and manganese to personganese cannot be far in the future.

The names of the newest transactinide elements pose further problems, as save exceptions Element 109, meitnerium named after Lisa Meitner, and Element 106, seaborgium named after the American Glenn T. Seaborg, they bear names of dead white European men. Element 96, curium, is another exception in that both Pierre and Marie Curie may be associated with that name.

It is uncertain what sex the Greek muse Urania might be, given John Milton's Paradise Lost: "For spirits, when they like Can either sex assume, or both".

The transactinides include:

92	U, **Uranium** (♀)	103	Lr, Lawrencium (♂)
93	Np, Neptunium (♂)	104	Rf, Rutherfordium (♂)
94	Pu, Plutonium (♂)	105	Ha, Hahnium (♂)
95	Am, Americium	106	Sg, Seaborgium (♂)
96	Cu, **Curium** (♂/♀)	107	Bh, Bohrium (♂)
97	Bk, Berkelium	108	Hs, Hassium
98	Cf, Californium	109	Mt, **Meitnerium** (♀)
99	Es, Einsteinium (♂)	110	Unnamed (*eka*Pt)
100	Fm, Fermium (♂)	111	Unnamed (*eka*Au)
101	Md, Mendelevium (♂)	112	Unnamed (*eka*Hg)
102	No, Nobelium (♂)	114	Unnamed (*eka*Pb)

We see two more transgressions here: Element 101, mendelevium (Md) named after a man and possessing the intolerable "men" syllable too clearly emphasizes the male sex component; persondelevium is preferred. By extension, berkelium (Bk) may represent not Berkeley,CA, but birkie, a pert young *fellow*, thus sexist. It depends on how you pronounce words.

Perhaps the yet unnamed and yet-to-be-discovered superheavy nuclei can be named more sensitively. The older *eka* prefix nomenclature will not do; nor shall Latin terminology "ununnilium" for Element 110, "unununium" for Element 111, etc. We note that at least one whole atom each of Elements $^{287}114$ and $^{289}114$ has been recently synthesized by cold fusion reactions, but claims of syntheses of Elements $^{289}116$ and $^{293}118$ have been since retracted.[12,34]

Use of the word "gay" now *de rigueur* for homosexual, now replacing "queer", came into media use in the early 1950s. The *New York Times* allowed the use only after 1986. The companion, alternative life style word "straight" for normal male sex behavior is also now infiltrating the literature. We note the opposite term "kinky" cannot be used, as this might be pejorative. Obviously, such words as fag, fruit, queer, dyke, and the like are proscribed.

As part of my hammering on the meaning of words thesis I regularly impudently told our students "On occasion I am gay". Do you really know what I mean? Another wicked test not used in class but one again revealing the power of words I have perhaps unkindly remarked to mixed assemblages "When women have oral intercourse with men they use fricatives". This simple sentence appears to stun or embarrass some, to confuse others, to be nonsense to yet others. I recall learning about fricatives and labials in elementary school, perhaps fourth or fifth grade, but that lesson may no longer exist in public schools. My latest use of this insulting test sentence received dismayed silence from a group of well-educated adults.

In the academic genealogy of the author there is the

French chemist Joseph Louis Gay-Lussac (1778-1850). Must we regard him as homosexual!

In addition to fear arising from innocent misuse of official terminology there is fear of litigation arising from disclosure of science misconduct and fraud. Those associated with miscreants may not be able to acknowledge the matter or publish a retraction or apology, as journal editors now fear lawsuits brought by those whose misconduct created the problem. For example, several coauthors of psychologist Stephen Breuning, convicted of criminal fraud, have tried unsuccessfully to get retractions of their published papers with Breuning, but lawyers for the journal publishers blocked such actions.

3. Effects on Biomedical Sciences

Standard English now the universal language of modern science has served science well for centuries. Intrusions of modern foolishness and many practices described here appear to be harmless; the mature scientist and aware student should have little problem in detecting and discounting items heavy with these flaws. Other practices, external political influences, Newspeak, and PC are insidious and impact science more seriously. Freedom of action is affected; exact meaning is more difficultly expressed; progress may be impeded by activism. Replacement of standard English by a variety of Newspeak or PC shall not serve science at all.

Besides error and confusion, matters of consequence that will be corrected shortly by others, in the extreme case events might follow the sequence: (1) A thesis is announced as being so; (2) Preliminary and inadequate evidence is adduced in favor; (3) Moral, ethical, and emotional disputation is advanced in support; (4) Personal and political advantage is perceived by adherents; (5) True believers emerge from cover; (6) Those who could never create an idea are able to exploit new ideas of others; (7) The bandwagon rolls; an agenda is created by insiders and adherents; (8) No discussion of issues not supporting the agenda; (9) Suppression of alternative concepts; rejection of data not in support; (10) Federal government intervention and funding.

These properties resemble those of "pathological science", "the science of things that aren't so", as defined by Irving Langmuir (1881-1957) in 1953 (Nobel Prize laureate in Chemistry, 1932): (1) The maximum effect observed is produced by a causative agent of barely detectable intensity. The magnitude of the effect is substantially independent of the intensity of the cause. (2) The effect is of a magnitude that remains close to the limit of detectability. Many measurements are necessary because of the low statistical significance of the results. (3) Great accuracy is claimed. (4) A fantastic theory, contrary to experience, is put forth. (5) Criticisms are met with *ad hoc* excuses. (6) The ratio of supporters to critics rises to near 50%, then falls to oblivion.

We see these features in the Polywater and "cold fusion" excursions. Elements may be perceived in other topics discussed hereinafter. In our defense, there is now the *Quarterly Review of Doublespeak*, Committee on Public Doublespeak, National Council of Teachers of English, Urbana,IL. And then there is Ebonics.

CHAPTER 5. MODERN REVOLUTIONARY AND MIRACLE SCIENCE

People Who Can Define Are Masters
- Stokley Carmichael

$\mathbf{A}$mong other burdens of modern biomedical science are the continuing news media inordinate announcements of science and medical "breakthroughs", with promises of great benefits to society. Although the published science journal articles of our interest are far from these news items, they are at times very much alike in that they may mislead, and news media stories may raise false hopes of remedies that do not yet exist.

A career among some of these events is helpful in possibly recognizing good from bad, certain from uncertain, utter foolishness, and media hyperactivity. Few indeed are newspaper, magazine, radio, or television reports of science that are correct and that satisfy curiosity with adequate details.

One must remember that journalism activities are designed to sell newspapers and advertisements for products and services. William Osler admonished us: "Believe nothing you see in the newspapers. They have done more to create dissatisfaction than all other agencies. If you see anything in them you know is true, begin to doubt it at once".

One is amused by announcements of revolutionary science, of miracle science, involving all manner of developments. We see superconductivity at high temperatures, "cold fusion", the "gay gene", the latest cancer cure, Viagra®, and other remarkable breakthroughs, some valid discoveries, many of uncertainty, some obviously foolish, some just wrong!

The now notorious faking of news stories in

newspapers, broadcast media, and books as well creation of entertainment in the guise of news is result of policy, that news must be entertaining but not necessarily accurate. Both sloppy or nonexistent editorial oversight or deliberate bias seems involved. However, a marked rise in corrections and errata in science journals dealing with minor editorial matters tells us of the current lowered standards in science as well. Both the public and scientists are now more skeptical of whatever new information comes their way.

Necessarily it is important to examine newsworthy items for validity before accepting them as so. Continuing evaluation of announced or confirmed science discoveries is essential, is expected, is required. Experimental data may be flawed or sound, interpretations drawn therefrom sound, inadequate, or flawed. Mere newsworthiness is not enough!

Moreover, besides the remarks of the enemies of science, the social constructionists, and postmodernists of Chapter 1, additionally the issue of science fraud or misconduct is now newsworthy. It is fashionable to write books dealing with such matters.[1] Some of these books are by journalists uneducated in science beyond generalities; other books delve into misconduct extensively. The refutation by William Bateson of evidence of Paul Kammerer for Lamarckian inheritance of nuptial pads in the midwife toad, with later claims by Hans Przibram and Kingsley Noble of fraud by Kammerer, is a classic example.[2]

Another case discloses character assassination as well. Claims of Franz Moewus that his *Chlamydomonas* mutant cells treated with rutin were no longer able to copulate are now recognized as being unreproducible, but Moewus worked in Germany 1931-1959 and was maligned as a Nazi (not so) by American Jewish scientists T. M. Sonneborn, J. Lederberg, and R. Sager. Moewus is presently discredited, his work dismissed.[3]

1. Older Cases

Moreover, it is now *de rigueur* to review older issues of science as if they were instances of misconduct or fraud, not just of error. The early discovery of the Laws of Nature once an obsession with scientists concerned lest all laws be discovered by others has now received such attention. Although searches for undiscovered natural laws no longer dominate science, it has become a pastime of some to debase the men who discovered fundamental natural laws, as if the law must be repealed. Despite his motto *"Hypotheses non fingo"* suggestion that Isaac Newton adjusted data has emerged! Nonetheless, the Law of Gravitational Attraction yet holds.

Of particular appeal is a reexamination of accounts of the early 19th century discoveries of fundamental laws of chemistry. One of these early fundamental discoveries was John Dalton's 1803 Law of Combining Proportions known to every student of general chemistry. Dalton noted that one reaction of nitric oxide (NO) and oxygen (O_2) needed twice the volume of NO as the other and claimed that his results

$$4NO + O_2 \rightarrow 2N_2O_3$$

$$2NO + O_2 \rightarrow 2NO_2 \rightarrow N_2O_4$$

confirmed the atomic theory of matter. Revisionists seeking error or fraud claimed Dalton faked data, but modern laboratory repetition of his experiments the way he did them confirm his data and his conclusions.[4]

The Law of Dulong and Petit of 1819 has also been questioned as a matter of fraud. The law states that specific heat times atomic weight is a constant.[5] However, the atomic weights for cobalt and tellurium used by Dulong and Petit were not correct, yet specific heats were obtained to meet the law. Their atomic weight of cobalt was two thirds today's value, that of tellurium one-half, leading them to use specific heats 1.5 greater for cobalt, twice for tellurium, to obtain results consistent with the Law. Dulong and Petit had reported their results within one week of their discovery, as discovery of new fundamental laws of chemistry was then the rage. The work was recently criticized as a matter of wishful thinking and of having such strong expectations that the data were adjusted to fit the Law. Yet more recent review of the matter indicates that Dulong and Petit were not in error on this point.

Those making other fundamental discoveries have been denounced as well. Friedrich August Kekulé's discovery in 1865 of the structure of the aromatic benzene ring system is variously disputed as a dream in which a snake swallowing its tail led to the cyclohexatriene ring structure. Kekulé is accused of scientific misconduct in that he must have been aware of previously proposed ring structures (but not his cyclohexatriene structure) of Austrian chemist Joseph Loschmidt and German chemist Albert Ladenburg. If Kekulé dreamed the matter there would be no need to cite prior reports; therefore misconduct! The charges are refuted.[6,7]

Recent review of the laboratory notebooks of Louis Pasteur has also led to charges of misconduct against Pasteur. Among the claims against him are:[8] (1) Citation of early but not later work of Auguste Laurent, whose earlier work influenced Pasteur's studies of optical isomerism; (2) Dubious ethics in early use of rabies vaccination; (3) Surreptitious use of the chemical attenuation method of Jean-Joseph Henry Toussaint without creditation in preparing his anthrax vaccine; (4) Debates with Felix-Archimede Pouchet on spontaneous generation 1884 in which he failed to approach a question without preconceived ideas and to disprove any opposing hypotheses; (5) Promotion of the myth of himself as scientific hero above reproach.

The classic studies of Johann Gregor Mendel with sweet peas leading to modern genetics have been criticized as being too good! Statistical analysis of his data in 1936 suggested chance of such results as 1:10,000. But yet Mendel was correct in his conclusions, whatever data he collected,[9] and the science of genetics flourishes nonetheless.

Other early 20th century items are recognized as

unreproducible, fancy, misconduct, hoax, or fraud. Among these are the N-rays of René Blondlot, 1903, that penetrated matter X-rays did not. The phenomenon was confirmed by others but was suspect. In one demonstration by Blondlot in a darkened room physicist Robert W. Wood surreptitiously removed a required prism from the apparatus, but Blondlot continued to detect N-rays.[10] The matter has long been concluded as involving observer bias, thus is without merit.[11]

More recent critical claims abound. Robert Andrews Millikan's famous oil drop experiment measuring the charge of the electron, for which he received the Nobel Prize, has been questioned by newspaper reporters whose understanding of experimental science is imperfect.[12] The sex studies in Samoa of Margaret Mead have been criticized in that she apparently took the stories of Samoans on faith, did not conduct original field work, but rather relied on what she had been told as so. Upon criticism that one of her conclusions was not so she responded "If it isn't, it ought to be"![13]

The revisionist business extends to other topics as well. Now we are told that Marco Polo never made it to China.[14] What is one to believe, to accept as so, given the problems of newsworthiness and the deficiencies of the news media.

The introduction of truly original ideas into science has always been fraught with doubts, rejections, even perils. What is believed today was ignored or rejected just yesterday. The now accepted concept of plate tectonics was pure daftness not too long ago. The demise of the Cretaceous dinosaurs following an asteroid hit near the Yucatan once a wild notion is now better understood. I regularly told our students that if you have a genuinely original idea, one that departs from present knowledge of the science master, professor, investigator, or pundit, prepare to have the idea rejected. If I have not heard of it, I being a funded investigator, how can your naïve notion have any merit.

Political intrigue also influences development of new ideas. The claim that Clovis Man (*ca.* 11,000 ago) was the earliest known human in North America has influenced examination of evidence of yet earlier human habitation in the Americas. Negroid features of the famous Olmec heads suggest African origins, and different Asian influxes are suggested by recent discoveries in South America. The 9,400-year old Kennewick Man skeleton recently discovered near Kennewick,WA, has features not those of American Indians, features perhaps of Ainu.[15] We recognize now that Clovis Man may not be the earliest human in America.[16]

The Kennewick Man discovery appears to be of utmost archeological importance, but the skeleton is in the protective custody of the Department of Interior, and access to the bones is denied to archeologists. Scientific examination of the bones is considered to be irreverent by government fiat under current PC multiculturalism concepts. Moreover, under the Native American Graves Protection and Repatriation Act of 1990, all ancient human skeletons, bones, and artifacts must be given to some regional Indian tribe near the archeological find for reburial. Whether the appointed local tribe inhabited the region in ancient times is not relevant; whether the skeleton be that of an American Indian is of no importance.

Meanwhile, several bones are missing from inventory, and the archeological site itself has been destroyed by the U.S. Army Corps of Engineers dumping 500 tons of rock and dirt "to protect and stabilize" the site, thereby also to deny further archeology exploration of what appears to be a very important site. The federal government argues otherwise.[17]

2. Which Are Real?

There are genuine discoveries that do revolutionize science, without journalists' attention. The question must arise: which new science event however announced is real, which questionable, which not genuine? Leaving out of consideration such items as religion and superstition, the "paranormal" and "pseudoscience", hoaxes, and foolishness recognizable on their faces, one yet must deal with the gullibility of the general public and journalists, but also with scientists involved in each miracle event.

Here is a list of a dozen modern newsworthy events of science. A much longer list could be prepared. Which are bogus despite claims to the contrary by the discoverers, advocates, and wishful thinkers. Which are error or mere foolishness; which are genuine discoveries of fundamental importance.

(1) Chemical elements are readily transmuted one into another.

(2) Scotophobin transplants the memory of trained animals into naïve ones.

(3) The inert noble gases helium, neon, etc. actually form stable chemical compounds.

(4) Water has a memory of past treatment at infinite dilution (to 10^{120} dilution).

(5) A previously undiscovered third elemental form of carbon neither graphite nor diamond is the C_{60}-molecule buckminsterfullerene discovered in Texas and now proclaimed the official molecule of the State of Texas!

(6) Water forms a polymer called Polywater that threatens ordinary life as we know it.

(7) Enzymes made of D-α-amino acids are active on the enantiomers of the substrates acted on by naturally occurring enzymes composed of L-α-amino acids.

(8) Ancient artifacts such as the pre-Columbian Vinland Map showing the outline of the Labrador coastline and the Shroud of Turin revered as the true burial shroud of Jesus Christ are proven genuine articles.

(9) Electric power superconductivity occurs well above liquid helium temperatures.

(10) Large unrecognized man-like animals yet exist on the land, large dinosaur-like animals in the waters.

(11) A "RNA World" in which life based on ribose pyranoside- rather than deoxyribose furanoside-nucleic acids preceded our present world wherein life is based on DNA.

(12) Heat useful to generate electricity is released from simple electrochemical cells by a process of nuclear fusion termed "cold fusion".

Even a precursory consideration of these items reveals that every other one is a genuine discovery, the others bogus, devoid of reliable science support. But is it the even-numbered or odd-numbered items that be sound. A brief discussion of several of these items is justified.

Transmutation of the Elements. We see the transmutation of elements as a regular event today, by irradiating a target isotope with an appropriate projectile, including the cold fusion synthesis of superheavy transactinide elements discussed in Chapter 4. However, who would understand or accept as fact the first reported synthesis of a new transuranium element, Element 93 now neptunium (Np), by neutron capture in the reaction:

$$^{239}U + n \rightarrow {}^{239}Np + \beta^-$$

Fission of uranium was understood at the time, but the capture of a neutron in "hot fusion" (distinguished from cold fusion) to form a new element was a novel synthesis not previously known.[18]

More recently the age-old alchemist dream of transmutation of inexpensive elements into valuable gold has resurfaced as an outgrowth of the "cold fusion" mania involving formation of ^{4}He and ^{3}He isotopes of helium (oops, itlium) from deuterium (^{2}H$_2$), *vide infra*! Distinguished Professor John O'M. Bockris of Texas A & M University who initially supported "cold fusion" generation of ^{3}H$_2$, has now proposed transmutation of the elements, to make gold from mercury. He also proposed a 4-body (!)

nuclear reaction for transmutation of carbon to iron:

$$2\,^{12}C + 2\,^{16}O \rightarrow\ ^{54}Ni + 2n$$

with subsequent decay of ^{54}Ni to ^{54}Fe! Public press coverage but no published science literature is now the case.

Scotophobin (The Ungar Case). A pentadecapeptide, scotophobin, recovered from brain of mice trained to avoid the dark, injected into untrained mice caused those mice to avoid the dark also. Several other laboratories are said to have confirmed these studies, but severe technical questions remain regarding the full story.[19-21] It staggers the imagination that a specific pentadecapeptide could be detected and isolated using the bioassay methods described. Scotophobin no longer is a subject of science investigation.

Reactive Inert Gases. The age-old acceptance of inertness of the noble gases was shattered in 1962 by Neil Bartlett with his discovery that platinum hexafluoride PtF_6 oxidizes xenon to give the salt $Xe^+PtF_6^-$.

Water with Memory (The Benveniste Case). The authors suggest that an aqueous antibody solution diluted to one part in 10^{120} retained the ability to evoke biological responses in polymorphonuclear basophils despite the improbability that any antibody molecule would be in the diluted test solution. The authors propose that water itself retain memory of the antibody, acting as a template.[22] The manuscript was instantly controversial, as antibodies cannot be detected at dilutions greater than 10^{-9} M. Moreover, by the Avogadro number only one antibody molecule could be present in 10^{-14} M dilutions, yet the more diluted solutions exerted the biological effect! *Nature* published the paper only after two years of review because the referees could find no obvious sources of experimental error! At the end of the text the editor placed an extraordinary advisory describing the work as incredulous and promising additional investigation.

John Maddox editor of *Nature*, magician Randi "The Amazing", and NIH anti-fraud expert Walter W. Stewart visited J. Benveniste's laboratory and concluded that results were delusion and unreproducible.[23] Benveniste in turn criticized the critics, likening them to the Salem witch-hunters and McCarthy-like prosecutors.[24] Clearly sending a magician to discover any slight-of-hand operations and the NIH fraud specialist who had been particularly critical of the manuscript prior to publication were measures approaching those of the Inquisition. However the work is further discounted by later experimental work,[25] and the controversy has continued.[26] Benveniste apparently lost his INSERM laboratory over the matter. Water memory no longer receives the attention of scientists.

Buckminsterfullerenes. A genuine discovery of a third form of elemental carbon, for which the Nobel Prize in Chemistry was awarded 1996 to Robert F. Curl and Richard E. Smalley, Rice University, and Harold W. Kroto, University of Sussex. The C_{60}, C_{70}, and related three-dimensional carbon molecules are known as buckminsterfullerenes, or just fullerenes, named after architect Buckminster Fuller's geodesic dome at the 1967 World Fair at Montreal,PQ.[27] The icosahedron C_{60} molecules must have been in the world forever but were discovered as such only in 1985. The three-dimensional structure of the C_{60} fullerene is said to have come from Smalley making a paper model in his kitchen, much like the discoveries of the structure of benzene by Kekulé and of the DNA double helix by Watson and Crick.

This discovery is of interest for our course, as initial claims of a newly discovered third form of elemental carbon that has been around since carbon first formed might have been in error. Surely the naïve scientist, if aware about such matters, could not be certain. The explosion in studies with fullerenes has yet to abate. Lest fullerenes be deemed too far outside of modern molecular biology, consider that fullerene derivatives inhibit HIV-1 protease, a matter of substantial biomedical interest.[28] A fullerene carboxylic acid is cytotoxic to cultured cells following irradiation with visible light. Visible light excites the fullerene to triplet state that then converts ground-state dioxygen $^3O_2$ to $^1O_2$ that is the toxic agent destroying DNA.[29] Synthesis of half the molecule $C_{30}H_{12}$ with hydrogen atoms about the rim has been achieved,[30] opening yet other directions for investigations.

Polywater. Polywater conceived to be a polymer of water is discussed earlier in detail (Chapter 2). The episode is a perfect example of bogus science that is nonetheless newsworthy.

D-Enzymes. As required by Nature, naturally occurring L-enzymes composed of L-α-amino acids act on their natural substrates to give conventional products, and synthetic D-enzymes of the exact same composition but of D-α-amino acids must act on enantiomeric substrates to give the enantiomeric product.[31] Such departure from mainstream modern science is also exemplified by recent studies of ß-peptides composed of ß-amino acids in place of the conventional α-amino acids and of peptoids, achiral N-substituted glycines, all adopting well-defined helical conformations.[32]

The Vinland Map and the Shroud of Turin. Among cases where science evaluation of specific historical artifacts is necessary are those of the Vinland Map and the Shroud of Turin, both highly controversial matters that have stumped a large coterie of experts. The Vinland Map discovered 1957 in a bound book, The Tartar Relation, of the travels of John de Plano Carpini to Mongolia 1245-1247, purported to be from 1440, shows Europe, Iceland, Greenland, and Vinland, thus the North American coast! Despite much argument supporting the map as authentic, polarized light microscopy and spectral examinations identified the modern pigment titanium white (TiO_2, anatase) on the inked map lines, thus indicating a modern fake dating from about 1920-1921. However, other particle induced X-ray emission data suggest very low levels of titanium (10 ng/sq. cm., 0.00062% by weight).[33-35] It remains uncertain whether the Vinland Map be a hoax in the class with Piltdown Man or whether other explanation may be discovered for its provenance.

The Shroud of Turin poses a deeper, more perplexing problem, as the Roman Church venerates the cloth as the burial shroud of Jesus Christ. Relics of the Roman Church generally have not received the attention paid the Shroud of Turin, as pieces of the true cross seem unreal, as does also the claim by Bertrand Russell that a church in Germany has the foreskin of Jesus Christ. The Shroud known only from 1355 was first examined in 1973, with conclusion that the image of a crucified man was an artist's creation.

The Roman Church later convinced the investigators of their error, and they recanted. More recent examinations in 1988 by radiocarbon dating methods, including accelerator mass spectrometry, gave dates to 1325-1361, and microscope examination of material on sticky tapes taken from the cloth revealed collagen tempura paint medium and pigments red ochre and vermillion present only in image areas. No blood or body fluid residue was found. It was concluded that a medieval artist painted the Shroud about 1355, it to become a relic in a Roman Church 1356.[34]

There are continuing claims of the Shroud's authenticity in pseudoscience articles and books.[36] Dr. Max Frei, the Zürich criminologist who validated the bogus Hitler diaries for the magazine *Stern*, claimed pollen grains from a Palestine flower are present on the Shroud and has new evidence of blood and DNA on the shroud, together with an age 1000 years older than suspected. Of course, pollen grains could be placed on the Shroud at any time, and the dead do not bleed.

My own experience with this matter occurred when I was asked to referee a manuscript for a forensic pathology journal. I was faced with the authors' attempt to authenticate the age of the Shroud of Turin by comparing it as authentic reference material to a 3rd-4th century C.E. cloth (of established age) as an item of questionable age, an example of switched meanings. Comparison of an item of uncertain provenance but deemed of established age with a correctly dated item of antiquity is rank error if not misconduct or fraud. The medieval artist who created the Shroud for the church cannot be guilty of error, hoax, misconduct, or fraud, but what may we conclude of the political and religious pressures to adopt official viewpoints clearly in error.

Superconductivity. Superconductivity at temperatures above that of liquid helium described more fully in Chapter 4 is a real phenomenon, and funding is good and continuing. Recall that Nobel Prizes were awarded for the discoverers.

Undiscovered Large Animals. There are unconfirmed reports of yet undiscovered large man-like animals such as the Yeti (Abominable Snowman) and Bigfoot (Sasquatch) that exist in the wilds. Reports of Yeti have surfaced since 1832, with tracks in snow, a scalp of uncertain animal origin, and a few questionable testimonials the only artifacts in support. A Tibetan lama suggests Yetis will not be seen, "as yetis do not like the smell of foreigners". There is also the reptilian monster "Nessie" in Loch Ness in Scotland.

The RNA World. In the speculative "RNA World" it is proposed that the origin of life on earth may have implicated RNA over DNA as a most important first carrier of life information. The concept is supported by two Nobel Prize discoveries, that in Chemistry 1989 awarded to Thomas Cech and Sidney Altman for their independent discovery of catalysis by RNA and the discovery of reverse transcriptase enzymes that form DNA from RNA now an accepted matter. Additionally, there are the provocative questions of Albert Eschenmoser: why are natural nucleic acids derivatives of a pentose and not a hexose? Why ribose? Why ribofuranose and not ribopyranose?[37] He posed that hexose pyranoside structures (and not pentose furanoside structures) of the sugar moieties of nucleic acids be prebiological precursors of natural RNA. Examination of the properties of synthesized "structural alternatives" hexose pyranosyl (6' → 4') oligonucleotides with the (6' → 4') linkage analogous to the (5' → 3') linkage of pentose furanosyl (5' → 3') nucleic acids revealed that the Watson-Crick (antiparallel strand orientation) purine-pyrimidine pairing is stronger than in natural RNA. Suggestion was made that pyranosyl-RNA be a more selective pairing system than natural DNA and have a greater potential for constitutional self-assembly in a natural environment than natural RNA. Such inquiry shakes the world of the nonce molecular biologist. Could it be so?

"Cold Fusion" (The Pons and Fleischmann Case). As previously noted, the term "cold fusion" was first applied to the nuclear synthesis of high mass transactinide elements. However, the term is now more popularly applied to the sensational alleged release of excess energy in the form of heat from inexpensive electrochemical cells having a palladium metal lattice as cathode bathed with $^2H_2$.

Electrochemical "cold fusion", if so, must involve a nuclear fusion reaction that produces heavier nuclei, with ejection of neutrons, protons, or gamma radiation, with release of energy as heat to be used for generation of electric power. Two equally probable nuclear processes occur in the fusion of two ^2H nuclei. An initially formed excited helium $^4He*$ nucleus decomposes to helium ^3He and tritium (^3H), with release of energy:

$$2\,^2H \rightarrow\,^4He* + 23.85 \text{ MeV}$$
$$^4He* \rightarrow\,^3He + n + 3.25 \text{ MeV}$$
$$^4He* \rightarrow\,^3H + {}^1H + 4.03 \text{ MeV}$$

Heat, neutrons, ^3He, and ^3H were variously claimed to be among reaction products, the heat being far in excess over the nuclear reaction products. Such disparity lead to the concept of heat generated from a nuclear reaction, "cold fusion", as no chemical reaction could produce so much excess heat. As no known nuclear reaction accounted for the results, a "hitherto unknown nuclear process or processes" was advanced.

Despite much activity in support, it appears no

reproducible experiment has been reported that confirms genuine nuclear fusion. No balance among nuclear products, radiation, and energy with theoretical nuclear physics has been fashioned. Electrochemical cold fusion appears to be of the same class as the Polywater escapade and several biology madnesses.

The electrochemical cold fusion episode teaches us more than one sound lesson. The exact same term "cold fusion" applied to two distinct processes creates uncertainty or confusion. We must paraphrase William Shakespeare and ask "Under which meaning, Bezonian. Speak or die"! The episode is also a perfect example of "Name It And It's Yours"; a popular, fashionable, or cute name may make you a celebrity along with your named goods.

Another lesson lies in the means by which electrochemical cold fusion was given to the world. The announcement was made in a press conference of March 23,1989, by B. Stanley Pons and Martin Fleischmann, University of Utah, Salt Lake City,UT. This technique of garnering priority of discovery, public attention, and, presumably, research funding is anathema to scholars who accept peer review of their manuscripts as means of reducing error. However, there is a growing tendency to circumvent the review process and make announcements in press conferences.

The lure of instant notoriety even prompts some self-promoting entrepreneurs to use public relations organizations for such activities. The matter became so troublesome to the *New England Journal of Medicine* that by the Ingelfinger Rule (named after former editor Franz J. Ingelfinger) a press release before publication shall result in rejection of a paper.[38]

The press conference claims of Pons and Fleischmann set many laboratories to work immediately seeking to confirm results. Press coverage was intense; Pons and Fleischmann became instant celebrities. Mentions of Nobel Prizes were abundant. Frenetic activity around the world followed. The press conference announcement was said to be prompted by suspected competition from another group, that of Steven E. Jones, Brigham Young University, about to announce results.

Pons and Fleischmann did publish their manuscript "Electrochemically induced nuclear fusion of deuterium",[39] as did also Jones.[40] All manner of support materialized rapidly. Barely three weeks after the press conference, an April 12,1989, session of the American Chemical Society meeting in Dallas,TX, addressed the topic. The State of Utah granted $5 million for development; the National Cold Fusion Institute was established. National Science Foundation funds for a cold fusion archive at Cornell University were sought by July 1989. Theoretical support appeared rapidly;[41] the journal editor excused publication in that the paper gave new theoretical treatment, whatever the facts of experimental "cold fusion".

As is the practice for such burgeoning events (Polywater, the Spector fraud, the Fossel NMR test for cancer) specialist meetings and workshops developed. There was the Workshop on Cold Fusion Phenomena at Santa Fe,NM, May 1989, sponsored by the U.S. Department of Energy. However, questions emerged rapidly. Peer review of work was abandoned in the mad scramble of true believers seeking to prove and skeptics to disprove. By May 1989 the ranks of disbelievers outnumbered the true believers. A more cautious workshop name appeared in the Workshop on Anomalous Effects in Deuterated Materials, Washington City,DC, sponsored by the National Science Foundation October 1989. Experimental support of others did not appear; such remarks as "Cold Fusion Produces Heat But No Papers" did appear.[42]

As doubts became more prominent Pons and Fleischmann left Utah 1990 and were hard to find for a while, but the University of Utah planned on continuing cold research.[43] Many subsequent conferences and meetings have been held, some apparently contentious in nature, with advocates and disbelievers at odds. The National Cold Fusion Institute closed June 30,1991. Charges of fraud and delusion appeared at an American Physical Society meeting in Baltimore,MD, May 1992. U.S. Department of Energy funding did not happen. The Utah Cold Fusion Laboratory closed 1994. Pons and Fleischmann published another paper with less provocative title and offered only a disclaimer in explanation that chemical explanations for excess heat "must be excluded".[44]

All manner of criticisms appeared; controversial, even rancorous, remarks were published by supporters and by detractors. Typical remarks include those of D. R. O. Morrison, CERN, Switzerland: "It appears that the laws of physics apply everywhere in the universe... except in cold fusion experiments performed by believers" but also "the continued skepticism of Drs. Morrison and Huizenga has to be ignored.." appearing at the Third International Conference on Cold Fusion, at Nagoya, Japan, 1992.[45,46] *Nature* no longer accepts cold fusion manuscripts.

Skeptics described the work of Pons and Fleischmann at the University of Utah and of Jones at Brigham Young University with the pejorative term "The Utah Effect" referring to the 1972 unconfirmed claim of Edward Eyring, University of Utah, for an X-ray laser, for which no theory in support was advanced.

Jones an early advocate of cold fusion in 1989 now a skeptic and Lee D. Hansen published criticism of cold fusion work of Melvin H. Miles, Naval Air Warfare Center Weapons Division, China Lake,CA, who then claimed he had been accused of fraud. Criticisms back and forth ensued.[47] The controversy has passed on to Internet news group discussions.

Detection of supposed product ^{4}He in some experiments is criticized for want of measurements made on ambient laboratory air at the time of the experiment, thus no means of telling whether atmospheric ^{4}He had contaminated the cell. Likewise, ^{3}H as a product of ^{4}He decay might not arise from fusion, as ^{3}H is found in Pd samples that may have been used as a filter for hydrogen isotopes. Other similar technical objections abound.

A session of the American Chemical Society meeting at Anaheim,CA, in March 1995 included fourteen "cold fusion" talks that were later assigned poster status. The poster session was held in an overheated underground garage. Three posters were withdrawn, six were no-shows; only five posters were actually presented. In the same year a new "cold fusion" journal *Infinite Energy: Cold Fusion and New Energy Technology* began publication. International conferences on cold fusion continued despite the furor. The Sixth International Conference on Cold Fusion was held October 13-18,1996, at Hokaido, Japan, the Seventh International Conference on Cold Fusion (ICCF-7) at Vancouver,BC, in 1998!

Funding costs of these excursions have been great, as expectations of success and a never ending supply of cheap electricity were too tempting. Other costs include at least one scientist killed in an explosion while conducting cold fusion work at Stanford Research Institute, Menlo Park,CA, January 2,1992. Once their domestic funding failed, Pons and Fleischmann continued their work in Nice, France, funded by the Japanese Ministry of International Trade and Industry, but that support ended in 1997. Experimental results supporting "cold fusion" have been neither confirmed nor explained despite great effort. Four books have been published on the subject.[48]

One cannot escape comparison of the "cold fusion" experience with the notorious Polywater episode described in Chapter 2, where a similar frenzy of activity occurred, activity recognized as dubious science and discounted as such. We also note that B. V. Deryagin who excited the Polywater episode entered the "cold fusion" investigations with an explanation of how the process might occur.[49] "Cold fusion" studies have continued for almost a decade, thus longer than those of Polywater. Moreover, it appears "cold fusion" studies may be continuing among true believers despite being discounted by most. It seems both Polywater and "cold fusion" developed from error, self-delusion, wishful thinking, and ambitious career building and self-aggrandizement generously funded by governments.

Besides these dozen revolutionary items there are oodles of others, some of lasting concern, others but of transient interest. Recall the canals on Mars described by Percival Lowell (1855-1916),[50] abandoned only after the July 1965 flight of the Mariner spacecraft. More recent 1976 discovery in National Aeronautics and Space Administration (NASA) space photographs of Mars of the image of a colossal stone face has been proposed as evidence of advanced life on Mars, even of civilizations! The assertion "there is no longer room for reasonable doubt of the artificial origin of the face" has been passed.[51] Others view the same photographs and see nothing but natural erosion features. Similarly disputed is an interpretation of microscopic images found in meteorites of Martian origin as representing life forms, the nanobacteria discussed here shortly. Neither large nor small images evince life on Mars, now or in the past.

Announcements of newly discovered genes related to ever so many human characteristics are now frequently had. The current politically sensitive issue of homosexuality has led to studies seeking biological origins of the affliction. Differences in brain neuroanatomy[52] and in the Xq28 region of the male X-chromosome[53] of normal and homosexual males have been recorded, the latter evidence giving rise to the concept of a "gay gene" disposing males to homosexuality. Both studies have been used for political gain; both have been severely criticized for imperfections in design and conduct. The cadaver brains that were examined were from men who had died of AIDS. The "gay gene" researcher himself may be homosexual, and he is under investigation by the Office of Research Integrity (ORI) of the Department of Health and Human Services (HHS), suspected of selective presentation of data favoring his point. As an NCI scientist he has been ordered not to discuss the matter.[54] There has been no independent confirmation of the matter.[55]

Perhaps the most appealing miracles of biomedical science are those dealing with newly devised remedies for serious health afflictions. Periodically special preparations are advanced as cures of cancer and other devastating disorders. Appealing names Krebiozen, Laetrile (apricot pits), and Thymosin are given the preparations, but their chemical compositions are not reliably disclosed, nor are such preparations useful medical agents. Useful for self-promotion of advocates, perhaps.

The present increased awareness of the importance of inspired oxygen in health matters brings about promotion of Vitamin O to increase the oxygen supply to the brain, apparently outside of the established hemoglobin carrier. Vitamin O offered for sale by Rose Creek Health Products, Kettle Falls,WA, is a saline supplement containing "stabilized oxygen" that upon absorption is carried via blood to cells and tissues, there "to prevent disease and promote health" (the product appears to be a very dilute sodium hypochlorite solution).[56] Liquid (*sic*) oxygen is also being promoted in similar fashion, as good for you.

The biochemistry of molecular oxygen of inspired air is of obvious, increasing importance, as the potential toxicity of reactive oxygen species is well recognized. However, current attention is given to the role of protective antioxidants in blood, tissues, and foods, not to means of increasing oxygen supply to the brain.

The growing interest in "alternative medicine" opens new opportunities for both improvement in medical practice but also in false claims for treatments made for money, not for the patient's well being. Sales of diet supplements and other preparations made of natural ingredients for relief of all manner of ailments ("snake oil") is age old and still prevalent but outside scientific or medical investigation. Among other alternative medicine concepts is that of the "therapeutic touch" whereby a practitioner detects the human energy field around a patient, important for alternative medical practice. However, a six-year old girl fashioned a test resulting in rejection of this proposition as a useful therapeutic measure.[57] Shades of Francis Galton's statistical arguments against spiritualism and prayer.[58]

Of course, now the government has entered the scene, the recently created (1992) Office of Alternative Medicine and the Office of Dietary Supplements within the NIH to conduct clinical trials of the effectiveness of home remedies.[59] Perhaps Krebiozen and Laetrile will be resurrected for government approved clinical trials to cure cancer! Then, forget not the claim in India of human urine as a therapeutic beverage curing amoebic dysentery, kidney failure, cancer, and a few other ailments! In dealing with these items one is reminded of Shakespeare: "What potions have I drunk of Siren tears".

3. Name It and It's Yours

A principle closely associated with miracle science and very important to learn is that by attaching a name to a concept, event, or material object we make it so. Who can argue with a name of something in print; it must exist; it is just up to us to find it or accept it. There really must be a square-circle or a round-square because we can proclaim the name.

However, it is not easy to get some names into the public domain. We have tales of Albert Szent-György offering godnose or ignose for Vitamin C and of Leon Morgan offering pandemonium for americium (Element 95) and delirium for curium (Element 96). A colleague at Lederle Laboratories attempted to christen his discovery of a new medically useful tetracycline antibiotic "Duggarin" after R. Duggar who had discovered chlortetracycline. Impossible, by company fiat; no product can be named after a mere employee. Editors and administrators block such nonsense.

One can no longer name their baby.

In other humorous fashion triamcinolone had to have a trademark name for sales, and the Lederle Laboratories official assigned the task of creating such name came up with "Superb-one", the "one" meaning ketone. This name was immediately elided by the steroid chemists to "Super-Bone". Like "molecular" today among the biologists, "Super-Bone" became a recognition signal for those of us who worked on this material. Triamcinolone became the generic name, "tri" for a three-feature functionalization (Δ^1-double bond, 9α-fluoro, and 16α-hydroxyl groups) of hydrocortisone product, "amcin" for American Cyanamid Co. (Lederle), and "olone" for the hydroxyl and ketone functional groups. In odd circumstance surely based in ignorance of the "amcin" syllables, the "cinolone" part of the name was later used by other steroid manufacturers for their different products.

It is customary to name things after their originators, but the names need not survive. My own experience is of this sort. A test of the presence of $^1O_2$ in biological systems using cholesterol as probe was once innocuously named the "Smith test" by William A. Pryor in a monograph. A second such label, the "Smith rearrangement" dealing with epimerization of cholesterol 7-hydroperoxide may be found in print. Both names were never used further and have passed into well deserved esoteric obscurity. Thus is everlasting fame lost.

More significantly, if names are attached to an entity in error or unawareness of history the true discoverers may be lost as the younger wave of investigators unaware of the history take up the name. The Haber-Weiss reaction of continuing concern in oxygen biochemistry, now termed the metal-catalyzed Haber-Weiss reaction, implicated in the

$$O_2^- + H_2O_2 \rightarrow HO^. + HO^- + O_2$$

generation of hydroxyl radical in biological systems, is an embodiment of the same reaction posed earlier by F. Haber and Richard Willstätter.[60] In this case, Willstätter did not need the credit.

4. Yet Unresolved Matters

There are uncountable numbers of research discoveries that might pose major alterations in thought and action were they confirmed. Absent confirmation these items remain of uncertain meaning and value, their issues unresolved. Many items are of such narrow interest that they attract no attention, right or wrong, and rest outside of public controversy. Others posing difficulties are merely ignored out of kindness. Genuinely troublesome articles receive direct attention, whether to resolution of no, as we have seen.

Six selected issues are examined here, none to resolution, for different reasons. Error or misconduct is not at issue, although these factors may be present in some cases.

Endogenous Human Cardiotonic Steroids. The regulation of salt and water balance in mammals is associated with the adrenal mineralocorticoid aldosterone *inter alia*. However, there is a long lingering suspicion that some other agent also be implicated, and unidentified endogenous factors have been proposed as Na^+K^+-ATPase inhibiting natriuretic hormones, variously termed endogenous digitalis-like, ouabain-like, and digoxin-like factors, detected immunologically. Accordingly, the 1991 report of the presence of the specific plant cardiotonic steroid glycoside ouabain in human adrenal cortex, plasma, and urine received much attention among endocrinologists and steroid chemists.[61]

The structure assigned from mass spectral data could not exclude an isomeric steroid. The factor might or might not be ouabain. Either it is ouabain or it is something else. Name It and It's Yours.

More surprising were claims that the material be biosynthesized in the human adrenal cortex! Neither accumulation from dietary sources nor cross-contamination of tissues with ouabain surely present in adjacent laboratory space was considered. Were the human adrenal to biosynthesize ouabain, five hydroxylation reactions (at the 1ß-, 5ß-, 19-, 11α-, and 14ß-sites) as well as steroid side-chain alterations and conjugation with rhamnose, all unprecedented in mammalian metabolism, must occur. Furthermore, the inadvertent, undetected cross-contamination of biological samples with steroids from adjacent laboratory areas is likely, given the very low levels of component detected and the peculiar properties of ouabain to complex with borate ion from borosilicate glass used in operations![62] The matter remains unresolved.

Cancer Treatments. As a totally different class of uncertainty there are many claims of medical treatments of human ailments. Here the boundary between ignorance, error, misconduct, and fraud may be difficult to discern, and "Dubious Science" may be ascribed to discounted, suspicious work. In the early 1930s there were physicians John R. Brinkley with his goat gland treatment for aging men and Harry Hoxsey with herbal cures of cancer, both men seen as saviors at work or as charlatans. Both were long ago exiled to Mexico where they continued their operations. It is uncertain whether these practitioners actually believed in their treatments, whether ignorance and incompetence were the case, or whether there was merely fraud for money.

More recent examples are found in the unconventional approach of Judah Folkman, Children's Hospital, Boston, MA, for tumor regression involving restriction of blood supply to the tumor by treatment with two proteins endostatin and angiostatin. Failures of others to confirm Folkman's results were met with rejoinder that it takes time to learn how to conduct the work. It appears the uncertainty lay in the reproducible production of the proteins, but the difficulties seem to be overcome, as human clinical trials are underway.[63] We note that Folkman was previously involved in an article dealing with the effectiveness of interferon α-2a in reducing the size of hemangiomas, an article of many errors requiring a correction and an apology to readers: "We are embarrassed by our errors and regret them".[64]

A related case is that of Stanislaw Burzynski, operating his cancer clinic in Houston,TX, who advances cancer treatment with antineoplastins that ameliorate or cure certain cancers. Because his patients are living testimonials but not part of Food and Drug Administration tests for efficacy and safety, Burzynski's program has received two decades of attention from the FDA attempting to require formal clinical trials. He was found not guilty of violating a court order regarding distribution of his antineoplastins in 1997, but the uncertainty of efficacy remains.[65]

These two cases demonstrate two major difficulties, that of confirmation by independent investigators and that of federal government fiat demanding compliance with bureaucratic regulations.

Ancient DNA. Recent public clamor over science fiction recreation of Jurassic dinosaurs from their fossil DNA is now matched by claims of recovery and amplification of fossil arthropod DNA from 20-120 million year-old amber.[66] However, such results have not been reproduced by others,[67] and the same proviso about care necessary for such studies, just mentioned for the work of Folkman with endostatin and angiostatin, has been made. Moreover, a fungal origin of the DNA recovered has been suggested.[68] The matter remains uncertain.

Thomas Jefferson. The continuing history question whether President Thomas Jefferson sired children by his slave Sally Hemings has been tested by male Y-chromosome descent and DNA analyses, with public announcements that it was he that fathered the children. However, Thomas Jefferson had no sons, so DNA of descendants of his uncle Field Jefferson was analyzed. The DNA evidence merely supports a Jefferson family member (uncle Field Jefferson, brother Randolph Jefferson, or several nephews) not necessarily Thomas Jefferson as the father. But circumstantial evidence supports the public announcements, and the public demands definitive conclusions. So much for science.

Nanobacteria. Uncertainty is not confined to these biomedical cases. National Aeronautics and Space

Administration (NASA) investigators using scanning electron microscope images found in meteorite rocks from Mars have interpreted these items as evidence of microfossil forms of Martian "nanobacteria", thus extraterrestrial life forms![69] Others reject the interpretation and suggest from transmission electron microscopy data that these forms be nanowhiskers of magnetite formed nonbiologically.[70] Similar images found in earth rocks as well as in Martian meteorites have not been adequately explained. These images provide us with another example of "Name It and It's Yours"; by merely naming these forms "nanobacteria" the objects become genuine prokaryotic cells. However, there is the notion that they be new forms of life, neither prokaryotes nor eukaryotes.

It remains uncertain whether such images in meteorites and terrestrial rocks, biological samples, and water sources may represent living forms or be artifacts of some sort. By contrast, there is intense interest in structures of nanometer size, the "nanoworld", and nanoscience (study of the nanoworld) now occurs from two directions. From simple chemistry to self assembly of molecules into complex clusters there is approach to the macromolecules of life, thus of life itself. Then from the macroworld and miniaturization of components important to solid-state electronics technology there is approach to the same dimensions of the nonliving nanoworld. The two approaches promise to provide whole new vistas for exploration of these matters, possibly to cast light on the origins of life. Additionally, NASA has added to the search for extraterrestrial life by creating the NASA Astrobiology Institute.

Moreover, we see the prefix nano- now in many other contexts, much the same as "molecular". There are now many nanowords: nanodrug, nanoengineer, nanoscience, and nanoworld terms in use, perhaps for teeny-weeny-itty-bitty size items. There are also now journals *Nanotechnology* and *Nano Letters*!

Global Warming. Nobel Prizes have been awarded for

discovery of atmospheric chemistry implicated in influencing weather on earth. The chemistry is authentic; commercially used fluorocarbons appear to react in the upper atmosphere to destroy ozone and cause other changes affecting weather.

It is also so that combustion of fossil fuels on earth releases carbon dioxide (CO_2) to the atmosphere, CO_2 then potentially reducing loss of heat from the earth, causing the so-called "greenhouse effect" leading to global warming. As these items may affect both the amount of radiation reaching earth and global warming likely to affect polar ice and sea level, the topic is worthy of exploration in class.

As with so many extrapolations from established science to the rest of the world, there are claims that polar ice thawing is the result of global warming; others view the thaw as a naturally occurring periodic cycle. Arctic Ocean ice has thinned about one foot, to 6.5 feet thick, in the last decade or so and retreated in area much further north. These and other similar issues remain uncertain despite application of modern science methods. Scientists know how to live with uncertainty.[71]

CHAPTER 6. ERROR, HOAX, MISCONDUCT, FRAUD & ETHICS

All men are bad,
And in their badness reign
- William Shakespeare

In this chapter we approach the matter of error in its many forms. Honest error is a natural result of experimental science and cannot be avoided. We generally recognize honest error at its discovery and correction. The error of misconduct involving deliberate deceit is error of another sort, one of the character of the miscreant implicated, one that may be admitted upon discovery or that may not be confessed no matter the facts disclosed. It is this second kind of error that now afflicts biomedical science, dishonesty that must be examined carefully in order to limit or avoid it.

How is the scientist to tell, how distinguish honest error from corruption. We divide treatment of these topics into four categories: (1) Error, being genuine error where there should be none; honest mistakes, misunderstanding, misinterpretation of data; distinguished from ignorance, incompetence, carelessness; wishful thinking, religious fervor, dogma; delusion, quest for the Holy Grail and Nobel Prize. (2) Hoax, a deliberate act intended to deceive; tricks, jokes, pranks, or serious fraud. (3) Misconduct, involving data "massage", adjustments; experiment rigging, design to effect goal; self citation, citation only of important persons; courtesy authorship extended and accepted; plagiarism; misuse of privileged information (manuscripts for review, grant applications); but now by Government fiat: "serious deviation from accepted practice" (National Science Foundation) and "Other serious deviations" (White House Office of Science & Technology, 1996). (4) Fraud, involving data fabrication; deliberate misinterpretations.

As each of these categories may be construed as error or as misconduct, it is crucial to distinguish between these terms as to precise meaning lest additional error be had in treating these matters. In some cases it is not certain which term be appropriate; indeed, error and misconduct may occur together so as to make good judgments uncertain. Decisions without full knowledge may not be possible. Certainly, these topics lead us again into one of the profound features of science, that of uncertainty. We may not reach the truth; moreover, we have no assurance that there is truth. We can reduce error but may not eliminate uncertainty.

Note that criminal matters of financial fraud, embezzlement, theft, illegal conversion, and the like are not included here. Likewise, other criminal issues of slander, libel, racial/sexual harassment, abuse of personnel, etc. are beyond present scope. Moreover, these activities find adjustment in the criminal justice systems, not among scientists.

1. Error

Error occurs in all aspects of our lives; error in science is no different from error in other matters. We recall Oliver Cromwell's 1650 appeal to the Scottish Kirk "I beseech you, in the bowels of Christ, think it possible you may be mistaken". Error covers each topic of this chapter, as we regard misconduct and fraud as error first, error in proper conduct. Rather than do nothing, the scientist tempts error. It hurts to reread some of our papers published long ago only to see an error unrecognized at the time, an undiscovered gaucherie, or typographical error, in print forever, bearing our names. Work that is merely embarrassing because of its naïveté may not be so much error, as the experience may develop a mature scientist more effectively.

Error comes in many forms, from innocence and ignorance, from arrogance and stubbornness, and from incompetence in thought and action. Reduction of error must be our goal, with guidance of more exerienced scientists and from the original literature free from recognizable error. Reliance on textbook material and review articles for guidance, without thereafter consulting the original literature, may lead to serious error, as textbook and review information is only so good as the information available from the original literature, which may be in error.

Obviously, textbooks and reviews can present only what is known from original literature, even if in error. As example, the ß-oxidation of dietary important linoleic acid described in older textbooks involves a combination of dienoyl coenzyme A hydratase and D-3-hydroxy-enoyl coenzyme A epimerase, followed by further ß-oxidation steps. It is now recognized that the ß-oxidation of polyunsaturated fatty acids involves an NADPH-dependent double bond reductase followed by double bond isomerase and hydratase steps. Not just a minor detail.

Another sort of error now may creep into literature as new and fashionable concepts are formulated upon speculations from what is known. A case in point is that of the oxysterols, simple cholesterol oxidation products found *in vivo*, and their many suggested physiological roles, particularly in control of 3-hydroxy-3-methylglutaryl coenzyme A reductase of sterol biosynthesis. In eagerness to be associated with fashionable arrangements, the uncertain physiological relevance of oxysterols, speculative in nature, is oft not disclosed. Thus, there is better support for biomedical research grant applications, even if in error.

The typographical error is of little importance in most cases, but we must note that a one-letter error in the word now transforms it to not, with major changed meaning! Also, "age" but not "ape", "protomer" but not "promoter". There is also the social error, perhaps misspelling an author's name or omitting it, where a statement offends someone, hurts their feelings.

More important errors fall into three further categories, minor error detectable by careful reading of text, major error casting doubt on the validity of results obtained or on conclusions drawn, and confession of falsifications resulting in retraction of the paper.[1] Error has also been categorized as reputable error and as disreputable error! Reputable error results despite careful work; disreputable error results from carelessness and sloppiness.[2] Genuine error where there should be none occupied our major attention, as the literature is full of it.

The Error of Curiosity. With respect to error, the error of curiosity does not exist. It must not be concluded that selection of an unpromising avenue of investigation is error of the sort under scrutiny here. Although a selected topic may be a mistake, no funding available, no means of experimental study, no interest by other parties, etc., following a problem from intellectual curiosity is never an error. If the scholar pursues a topic in whatever manner possible, with whatever funding or support is available, there is no error. The error would develop were one to tackle a problem solely for the research funding available, the promise of fame and fortune, of winning the Nobel Prize. One may never gain fame and fortune; one can fail.

My experience in this matter may be instructive. In my salad days I undertook to study the air oxidation of cholesterol as a phenomenon. Pure cholesterol is odorless; cholesterol in its bottle develops a characteristic odor. Therein was an intellectual provocation. Paper chromatography now an obsolete method was available but not the vastly more effective chromatography methods of today. Return to the problem years later led to our current understanding of cholesterol autoxidation.

Notably, my first NIH research grant application dealt with cholesterol oxidation but was disapproved, funding denied. I was sent an explanation: it was known that such matters occurred by peroxide formation and the work could be done in less than one year anyway. Indeed, the work involved hydroperoxide formation, and by plastering the word "atherosclerosis" on grant applications almost thirty years of grant support became available.

It was the practice of a university committee on which I served to give pep-talks on research to medical students undertaking a brief research opportunity between terms. These were bright students interested in research, but the last year I was involved I inopportunely described my entry into the cholesterol oxidation work as one of curiosity over the odor developed in otherwise pure cholesterol samples. Dead silence. How could anyone with some degree of accomplishment care about such a matter. This is not what biomedical research should be.

Today it remains unsettled whether cholesterol oxidation products, the oxysterols, present in human tissues and blood plasma have meaningful influences on the etiology of human atheromas, but oxidized blood components are finally getting their share of serious attention by funded investigators. Was it error to examine the oxidation process in such detail that our descriptions have entered into the known background now of all who undertake related studies? Although a link to the human health disorder remains obscure, not one moment was misdirected in the fundamental studies supporting the present status.

Honest Mistakes, Misinterpretation of Data. Error also comes from misinterpretation of data. Examples abound in the biomedical literature, but I mention here just one of my own mistakes. It is current practice to assign chemical structures to all manner of compounds from NMR, mass spectral, and X-ray crystallographic data, all assignments subject to data interpretations. The older measures as combustion analyses for elemental composition, optical activity for stereochemistry and absolute configuration, etc., are not always accumulated today, nor is confirmation of structure by independent synthesis regularly employed. Most simple structure assignments from spectra are probably correct, but the number of assignments in error is unknown. In test cases, laborious total synthesis may be necessary, witness the examples of antitumor natural products himastatin and diazonamide A, the correct structures being established only after independent total syntheses.[3]

Our assignment of structure to an oxidized cholesterol derivative was based on a one proton NMR signal that was misattributed to an alcohol proton (OH) instead of to the correct hydroperoxide proton (OOH), our prize example of error necessitating considerable further work for our subsequent revision of structure.[4] The matter was thoroughly considered before publication, but the wrong choice was made, to our chagrin. There it is in the journals, forever.

Wishful Thinking, Delusion. The several cases claiming the presence of $^1O_2$ in biological systems discussed in Chapter 2 may not be incorrect despite use of nonspecific interceptors. Were $^1O_2$ present it would be intercepted, but so would all other reactive oxygen species. More recent work with the specific 1.27 μm emission band of $^1O_2$ suggests that perhaps very low levels of $^1O_2$ may be generated in some of these cases. Unless and until the previously reported experiments be repeated using the now accepted specific emission band for detection, the matter must remain uncertain whether there was simple error or wishful thinking.

Wishful thinking in the face of opposing evidence is obviously error, but a charge of hoax, misconduct, or fraud becomes difficult to press. The previously discussed oxygen biochemistry (Zepp *et al.*; Paschen and Weser; Rosen and Klebanoff; Chapter 2) displays the error of wishful thinking coupled with ignorance of background.

Other prominent instances of error promoted by wishful thinking include the Vinland Map and the Shroud of Turin previously discussed in Chapter 5. In both these cases overly enthusiastic support of a thesis of early pre-Columbus world discovery and of the religious matter has intruded into sound science examinations of the artifacts. It remains to be settled whether the Vinland Map be a hoax; the Shroud of Turin merely appears to be among the relics of the Roman Church as an article of faith, science notwithstanding.

2. The Hoax

The hoax, an act intended to deceive others, may be a deliberate trick or a serious fraud. The hoax designed as a trick is not so much a problem in science as is deliberate, intentional fraud but still must be considered. There have been many hoaxes dealing with art works and historical documents. The unavailability of original documents serves well to expose such fakes. Items from remote inaccessible monasteries in Tibet or in lost copies of defunct newspapers tell us well enough of this sort of hoax or fraud. Such hoaxes are devised for financial gain by the hoaxers, but those of science seem to be to deceive true believers, those who take too fully at face value some proposition in controversy. The two cases that interest us here are almost 90 years apart and serve two different goals.

The Piltdown Man. The Piltdown man *Eoanthropus dawsoni* of 1912 known from a fossil human skull but ape jaw was hailed as the long sought simplistic "missing link" between man and apes. The hoax perpetrated to mock human evolution theory appears to be that of the then curator of zoology of the British Natural History Museum but remains unconfessed.[5]

The Sokal Hoax. Alan D. Sokal, Professor of Physics, New York University, submitted a manuscript with gibberish title "Transgressing the Boundaries: Toward a Transformative Hermeneutics of Quantum Gravity" that was accepted and published in a trendy "cultural studies" journal.[6] Hermeneutics is the study of methodological principles of interpretation!

Sokal sought to see if a manuscript liberally salted with nonsense postmodernism rhetoric dealing with the political "reconceptualization" of science that sounded good and flattered the editors' ideological conceptions would be published. The physics and math in the paper are nonsense, but politics and quasi-philosophy from "sci-tech studies" literature flattered both editors and contributors and was published. Among foolishness in the article were assertions that physical reality is merely a "linguistic construct", with a call for a new "postmodern and liberatory science"! Sokal later confessed the hoax that mocks modern idiocy related to science.[7]

Such occurrences pose little direct threat to modern scholarship, certainly not to science, save that they build toward the ultimate 1984. Nonetheless, Sokal was berated for his violation of the tradition of scholarly trust. Note that the one unconfessed hoax involved physical artifacts that could be examined, whereas the other was merely foolish verbiage that supported attacks on real science.

3. Misconduct

The issue of science misconduct is a more insidious matter, as definitions of misconduct remain uncertain, and federal government agencies now seek to enter the fray with their threats to devise means of handling misconduct if the scientists do not do so.[8] The government agencies have their own expanded definitions making misconduct approach being a federal crime.

The present treatment of error, hoax, misconduct, and fraud is thus arbitrary in some of its arrangements, as error is involved in all four topics, as also may be misconduct. Misconduct is ofttimes said to be falsification, fabrication, and plagiarism (acronym FFP) without mention of fraud, which may have a legal definition outside of science concerns. In paraphrased legal terms, fraud is an intentional perversion of truth for the purpose of inducing another to part with some valuable thing; misconduct is a transgression of an established and definite rule of action (synonyms misdemeanor, misdeed, misbehavior, delinquency, impropriety, mismanagement) but is not negligence or carelessness.[9] For present purposes fraud is treated as a separate item, it being here understood that fraud deals with falsification and fabrication of experimental results. Plagiarism is here treated as misconduct.

Misconduct Defined. There are many definitions of misconduct, each crafted to achieve some goal, with some purpose behind the definition.[10] Initial 1982 federal government definitions included mismanagement of funds, fraudulent or markedly irregular (!) practices in carrying out research procedures or handling research results, serious failures to comply with requirements governing the protection of human subjects and the welfare of laboratory animals, and serious failures to comply with any other conditions of an award such as the guidelines for research with recombinant DNA molecules. Implementation policies were shortly devised.[11]

Official definitions of misconduct have been modified, *vide infra.* Government proposals for misconduct investigations now extend to all research, whether federally funded or no, and seek to publicize all investigations, including names of the accused, whether proven or no. Somehow scientists and science must suffer their transgressions.

A simple threefold categorization of misconduct might be: misconduct in science (fabrication, falsification, or plagiarism), questionable research practices, and other misconduct. However, this simple definition appears to be defective to the federal government that has defined misconduct as including "other serious deviations", whatever that may mean. Another recent definition further emphasizes federal government interests: Misconduct is (1) fabrication, falsification, plagiarism, or deception in proposing, conducting, or reporting research results, or (2) material failure to comply with federal requirements that are uniquely related to the conduct of research.[12] Failure to comply with rules and regulations of federal agencies has become of major importance in their assessments of science misconduct.

Whereas the simple definitions can be understood, the devious, enigmatic "other serious deviations" feature has no limitations and presages added regulation of science by the federal government already controlling so much of biomedical research via funding. At the behest of Rep. John D. Dingell (Democrat from Michigan), then chairman of the U.S. House of Representatives Energy and Commerce Committee Subcommittee on Oversight and Investigations, the high-visibility Office of Scientific Integrity (OSI) was established within the NIH to oversee misconduct investigations by awardee institutions. Also established was the Office of Scientific Integrity Review (OSIR) under an assistant secretary of the Department of Health and Human Services (HHS) to monitor misconduct operations in research and to review scientific misconduct investigations "to ensure that documentation sufficiently supports recommendations".

This arrangement caused troubles between Rep. Dingell and then NIH director Bernadine P. Healy such that the operations were moved from NIH to HHS and renamed the Office of Research Integrity (ORI), divided into the two subdivisions Division of Research Integrity Assurance and Division of Policy.[13] The NSF handles misconduct cases in its Office of the Inspector General (OIG). Protection is accorded the whistleblower; anonymous accusations are investigated.[14] Falsely accused parties must suffer, by fiat.

These developments attest the real goal of such arrangements: the funded increase of federal bureaucracy, with all the organization charts and assigned territories, and the increase of federal government power over research.

As previously noted, in its simplicity misconduct may involve data "massage", adjustments of data to fit fancy, experiment rigging designed to effect goal, self citation, citation only of important persons, courtesy authorship extended and accepted, misuse of privileged information (manuscripts for review, grant applications), and plagiarism.

Misconduct also encompasses false claims of academic background and of earned academic degrees. The false claims on NIH research grant applications of Annmarie Surprenant, Ph.D., to have the M.D. degree from the

University of Illinois at Chicago and of Mark S. Chagnon, D.Sc., to have a M.S. degree in organic chemistry from MIT resulted in their exclusion from federal funding and from service on Public Health Service advisory boards for three years.[15] Misrepresentation of background is a common matter, as also are false claims of preliminary research results now so much required for serious consideration of research grant applications.

There are aspects of misconduct that are indeterminate as to what they be. Deliberate misinterpretation of data might be misconduct or fraud. Mistreatment or cavalier dismissal of personnel, nepotism, hiring family or personal friends, etc., may be misconduct or criminal matters. Science misconduct does not include fraud as such nor embezzlement, fiscal misuse of funds, theft of materials or services, or other criminal acts.

Official Definitions. Besides fabrication, falsification, and plagiarism in several definitions we see the additional insidious wording of government definitions of misconduct: "serious deviation from accepted practice" (National Science Foundation), "Other serious deviations" (White House Office of Science & Technology, OSTP, 1996), and "Other practices that seriously deviate from those that are commonly accepted within the scientific community" (Public Health Service). The growth of the problem is exemplified in the exact wording of several government definitions of misconduct:

Public Health Service (adopted 1989): "'Misconduct' or 'misconduct in science' means fabrication, falsification, plagiarism, or other practices that seriously deviate from those that are commonly accepted within the scientific community for proposing, conducting, or reporting research. It does not include honest error or honest differences in interpretations or judgments of data".[16] Here we see imperfect writing of the sort that entangles. Note the intrusive PC wording "scientific community" instead of "scientists". The pronoun "It" of the second sentence has no definite antecedent.

From the National Science Foundation (adopted 1991) we have "'Misconduct' means (1) fabrication, falsification, plagiarism, or other serious deviation from accepted practices in proposing, carrying out, or reporting results from activities funded by NSF; or (2) retaliation of any kind against a person who reported or provided information about suspected or alleged misconduct and who has not acted in bad faith". Here we see the beginnings of protection for the "whistleblower" who alleges misconduct but none for the falsely accused scientist.

Commission on Research Integrity (CRI), Public Health Service (proposed 1995): The CRI agency created by Congress in 1993 issued its 1995 report Implementation Proposals on Recommendations by the CRI with: "Research misconduct is significant misbehavior that improperly appropriates the intellectual property or contributions of others that intentionally impedes the progress of research, or that risks corruption of the scientific record or compromising the integrity of scientific practices. Such behaviors are unethical and unacceptable in proposing, conducting, or reporting research, or in the reviewing the proposals or research reports of others.

"Examples of research misconduct include, but are not limited to, the following:

"Misappropriation: An investigator or reviewer shall not intentionally or recklessly a. plagiarize; or b. make use of any information in breach of any duty of confidentiality associated with the review of any manuscript or grant application.

"Interference: An investigator or reviewer shall not intentionally and without authorization take or sequester or materially damage any research-related property of another.

"Misrepresentation: An investigator or reviewer shall not with intent to deceive, or in reckless disregard for the truth, (a) state or present a material or significant falsehood; or (b) omit a fact so that what is stated or presented as a whole states or presents a material or significant falsehood".

These government efforts were highly organized.[17] Kenneth J. Ryan, Harvard University Medical School, is chairman of CRI. There was an Implementation Group on Research Integrity and Misconduct (IGRIM) with panel chairman William Raub (Secretary of Health and Human Services Donna Shalala's science advisor). From the ubiquitous Office of Rulemaking of every federal Department comes notice in the *Federal Register* "Advanced Notice of Proposed Rulemaking" in which there is no proposal for nonadversarial resolution of misconduct cases. Only adversarial approach is to be used, with subpoena power for individuals and documents, and legalisms in definitions (*mens rea* = intent). We see movement away from science judgement of misconduct to legal judgments.

In the face of opposition over the "other practices" terminology OSTP has now proposed amended arguments that yet retain wordings presaging continued uncertainties regarding innovative approaches in new science: "Research misconduct is fabrication, falsification, or plagiarism in proposing, performing, or reviewing research or in reporting research results. A finding of research misconduct requires: A significant departure from accepted practices of the scientific community for maintaining the integrity of the research record. The misconduct be committed intentionally or knowingly or in reckless disregard of accepted practices. The allegation be proven by the preponderance of the evidence".[18]

Here we see elements of PC (scientific community, not scientists) but more insidiously of adversary legal procedures (preponderance of evidence, not common sense). One notes the increasing complexity of federal regulations as the federal bureaucracy intervenes, passing disturbingly from science fraud to science misconduct.[19] Moreover, criminal penalties for infractions are not that far in the future. Issue of confidentiality; public report on all cases, including accusations not judged or judged innocent. Whistleblowers are to be protected; false accusations have no consequences.

Under these restraints one wonders which recent expansions of science might not have occurred. In the biomedical sciences: discovery of prion infectious proteins, reverse transcriptase, and others that are a "serious deviation from accepted practice". In fundamental science, the mass of the neutrino![20] Each of these items was far from accepted practice; we are taught that proteins be not infectious, that DNA beget RNA beget protein, that the neutrino have no mass.

Charges of Misconduct. There is the ORI Newsletter issued quarterly by the Office of Research Integrity, U.S. Public Health Service, in which items of research integrity, ORI activities, and case summaries in which names are provided. The ORI maintains a "home page" on the World Wide Web of the Internet. Also, ORI now seeks to expand its authority to require that all persons involved in any way with research be given "Instruction in the Responsible Conduct of Research"! Principle investigators down to "anyone else involved in conducting research", including secretarial and general support staff must receive such instruction. It appears those who would conduct the instruction are now the ones to be instructed! [21]

There has been controversy about whether charges of misconduct must be reported before investigations are conducted, before any charges are proven. Some call for such disclosures; others recognize mistakes be possible in these matters. Two highly publicized cases exemplify the problems of political intrigue and character destruction.

Cancer surgeon Bernard Fisher, University of Pittsburgh, was charged with using fraudulent data of Roger Poisson, St. Luc Hospital, Montreal,PQ, in an international breast cancer lumpectomy study of the NCI National Surgical Adjuvant Breast and Bowel Project led by Fisher. He had informed NCI of the falsified data, and Poisson admitted he had falsified data, but Fisher was blamed and removed by NCI as head of the project and charged by ORI with using the fraudulent data of Poisson. The National

Library of Medicine took the extraordinary action of adding "science misconduct" alerts on all papers of the project, whether involving Poisson or no. The MEDLINE computer data were so marked, for all the world to see.

In their zeal to grasp attention, Rep. Dingell investigated the case in 1994, and feminist activist then Rep. Patricia Schroeder (Democrat from Colorado) decried "the scientific community in their ivory towers who take the public's money and then get offended when we ask them to be accountable". Her use of "scientific community" instead of the single word "scientists" revealed her commitment to PC but also to deception, as the alleged abuses were medical ones, not science matters. Thus flourishes inflammatory government seeking issues for political advantage!

After a two-year nine-month investigation Fisher was exonerated, but he had to file a lawsuit in federal court against NIH, NCI, ORI, and the University of Pittsburgh alleging his First Amendment Rights were compromised to have the government abuse adjusted. Fisher got an apology, praise, and a financial settlement.[22] Poisson admitted he had falsified data, was forced to retire from the University, but he continues practice in St. Luc Hospital.[23]

In a second example similar charges against oncologist David Plotkin for mismanagement of patient files in 1994 were made by a newspaper investigative reporter. Plotkin was exonerated 1995 by ORI in a press release! However, ORI investigations that do not support misconduct charges lead to exoneration but not to proclamation of innocence. So we see a public misconduct charge, a government investigation that does not establish misconduct, but then a public statement that does not proclaim innocence. Publication of unproven charges or of exoneration of charges merely taints the name and reputation of those charged.[24] Plotkin's question "How do I get my science reputation back?" is not answered by ORI.

The Baltimore Affair. The Baltimore Affair is one of the most notorious recent instances of misconduct, if not worse.

David Baltimore, Nobel Prize laureate in Physiology or Medicine (1975), colleague Thereza Imanishi-Kari (then at MIT), and others published a paper "Altered Repertoire of Endogenous Immunoglobulin Gene Expression in Transgenic Mice Containing a Rearranged Mu Heavy Chain Gene",[25] that was adversely criticized by junior colleague Margot O'Toole, who charged that Imanishi-Kari's work could not be repeated. Others reported similar difficulties.

Errors in the original paper were adjusted,[26] and four coauthors but not Imanishi-Kari retracted the paper.[27] She published two items supporting her position.[28] Personal differences among parties developed; O'Toole and Imanishi-Kari were at odds. Earlier charges of error became charges of misconduct by falsifying data. O'Toole became anathema. Baltimore initially protected Imanishi-Kari and discredited O'Toole, later distanced himself from Imanishi-Kari, apologized to O'Toole, and retracted the questionable paper but eventually supported Imanishi-Kari again.

The affair drew the attention of the NIH team of Walter W. Stewart and Ned Feder seeking science misconduct and of Rep. Dingell whose investigations 1988-1992 brought the matter to public scrutiny. Dingell had Imanishi-Kari's lab records examined by the U.S. Secret Service. The OSI investigated the problems. The OSI taken from NIH 1992 was renamed Office of Research Integrity (ORI) 1992 and reorganized into the Public Health Service, Department of Health and Human Services. Dingell blurred the distinction between error and fraud. As his inquisition lost its vigor he turned from fraud to misconduct and finally to expansion of government investigations "to ensure the continuing pre-eminence of American science".

Everyone lost something. Stewart and Feder eventually were reassigned to other work. Rep. Dingell lost his committee chairmanship in the 1994 national elections. Baltimore recently appointed in 1990 president of Rockefeller University resigned 1991 over the matter but was later (1997) appointed president of California Institute of Technology. The ORI initially concluded in 1994 that Imanishi-Kari was guilty of misconduct by fabricating data. She was to be barred for ten years from federal grant or contract funding and was demoted from assistant professor to research associate (she then at Tufts University). However, subsequent investigations failed to discover evidence of fraud but found only "sloppy" record keeping by Imanishi-Kari, whose laboratory records were in disarray. She was ultimately cleared of misconduct charges in 1996, regained her faculty appointment at Tufts University, and returned to her research. O'Toole as "whistleblower" was supported at first (1991), received a Cavallo Foundation, Cambridge,MA, $10,000 Prize for Moral Courage in 1992, but later had her credibility questioned. She had trouble finding subsequent appointments but is now employed in the biotechnology industry.[29]

Thus ended a ten-year odyssey from initial error to final disposition. Others have not tried to confirm the questioned results, and the matter is now laid to rest. Separate investigations by two universities, the OSI (NIH), the ORI (HHS), and a congressional committee had been made, the early ones examining possible errors, the later ones focusing on possible fraud! The greater damage was done once investigations by scientists passed into the hands of administration, the lawyers, and politicians.

The many contradictory aspects of the affair leave a feeling of uncertainty as to just what did occur. Failure to discover evidence of fraud is not exoneration of charges of fraud. The charges arose from questionable laboratory records that were not kept in good order. Experimental data are under strict control in most laboratories where the importance of good records is understood. Daily completion of necessary records, signed and dated, are regular activities, part of the job. The poor "sloppy" practices of Imanishi-Kari, while apparently not mounting to fraud, exemplify the potential problems of such inadequate operations, problems that become severe given circumstances of the Baltimore affair. The case is a perfect example of what can happen where poor lab practices, flourishing personal animosity, leadership failure of senior scientists, and politicized government investigations occur.

Yet other more ominous events reflect similar interlaboratory personnel conflicts. Poisoned coffee in the Robert Roeder laboratory at Rockefeller University[30] and ingestion of perhaps 820-1,200 mCi (8.0-12.7 rem) ^{32}P by a pregnant female investigator in the John Weinstein laboratory, NCI,[31] appear to be related to tensions about generation of research results. Tensions and conflicts have now resulted in suicides of graduate students at Harvard University.[32]

Plagiarism. Plagiarism, the taking of intellectual property of others without their knowledge or approval and without proper attribution, poses unsettling issues. The use by James D. Watson and Francis Crick of Rosalind Franklin's X-ray data without her knowledge or approval (discussed in Chapter 3) is not now regarded as plagiarism, as senior professor Maurice Wilkins, who made the data available, presumed he had that right. It is hard for me to agree about this matter, the utter disregard of one scientist's rights for the benefit of others.

Plagiarism is now considered misconduct but in other texts is considered to be fraud. Here again the line between misconduct and fraud may be obscure, but intent is a major factor in fraud. Outright copying of others' work can be readily detected and proven, as word by word comparisons reveal the matter. Plagiarism may involve frank theft of a work of others, copied verbatim into work of the thief. Whole articles may be copied and resubmitted to other journals with the plagiarist as author, as did Elias A. K. Alsabti who republished at least seven papers of others as his own.

Moreover, innocent or sloppy copying of too many parts of the work of others without attribution may be plagiarism. Established cases have been made against well-known national nonscientists Senator Joseph Biden, historians Stephen Ambrose, Alex Haley, and 1995 Pulitzer Prize winner Doris Kearns Goodwin. However, the more often encountered science cases involve misuse of privileged information incorporated into the miscreant's works. Two such notorious cases of unconfessed plagiarism demonstrate how otherwise successful scientists fall into serious misconduct. These cases involve misuse of privileged journal manuscripts and federal research grant application sent them for review.

In 1986 C. David Bridges, Baylor College of Medicine, Houston,TX, reviewed a manuscript of P. S. Bernstein, W. C. Law, and R. R. Rando dealing with enzymatic isomerization of all-*trans*-retinoids to 11-*cis*-retinoids later published in the *Proceedings of the National Academy of Sciences USA*[33] and published his own paper "The Visual Cycle Operates via an Isomerase Acting on All-*trans* Retinol in the Pigment Epithelium" shortly thereafter,[34] but without mention of the prior work of Bernstein *et al.*[35] Bridges' work quickly received attention; however, the similarity of the papers, with claim that Bernstein *et al.* have priority, was also noted directly.[36] Charges of misuse of the Bernstein *et al.* review manuscript were made, and what facts are known reveal that Bridges' work did not begin until after he received the Bernstein *et al.* manuscript for review.[37] As is the case so often in such disputes, the original laboratory notebooks of coauthor R. A. Alvarez cannot be located; instead prints of computer records were adduced. A NIH investigation found Bridges guilty of plagiarism; Bridges lost his NIH funding and left Baylor for Purdue University.

Leo A. Paquette, Professor of Chemistry, Ohio State University, was found guilty of plagiarizing material from a NIH grant proposal of Stephen F. Martin, University of Texas at Austin, reviewed by Paquette then chairing the NIH Study Section considering Martin's proposal. By story Paquette gave Martin's unfunded NIH application to an unnamed (shielded) postdoctoral fellow who then somehow incorporated the material into Paquette's own NIH application and later journal article of 1992.[38] Paquette was later charged with a second case of plagiarism in 1993 by the NSF Office of Inspector General. In Paquette's 1992 paper dealing with the chemistry it appears references cited by Martin were incorporated, even to the same misspelled words! It further appears he submitted falsified evidence refuting the misconduct charges. In a settlement of misconduct charges he was excluded from federal funding for two years, and NSF was not to make a misconduct charge against him. The example is that of the overly active leader of a research group of 30-40 persons, publishing 30-50 papers annually!

For balance there are other cases of plagiarism in medical matters, one being that of Vijay R. Soman, Yale University, accused of plagiarism and inventing patients in his studies of insulin action on monocytes. The case of Prof. Walter Frost, University of Tennessee, allowed U.S. Army and NASA officials to plagiarize his own work to get easy degrees under his supervision. The 1989 master thesis of NASA scientist Peggy Y. Potter is said to be identical to a paper published by Frost in 1987.[39] In return he obtained privileged government grants.

4. Fraud

The topic of confessed and unconfessed biomedical science fraud has already been described with several examples, including the notorious Mark Spector case, in Chapter 2. The present treatment builds from this prior material by adding new examples showing the diversity of frauds committed in recent decades.

As noted, error, misconduct, and fraud come in different guises not always easily distinguished. Error is the most frequently encountered but the least damaging to science, as error will be discovered. Misconduct already discussed is a more complex matter involving in large part research funding. In legal terms as well as everyday language fraud is criminal deception, use of false information with intent to benefit the deceiver. Intent is the key criterion, as simple error, no matter how egregious is neither fraud nor misconduct.

The aspect of intent may also differentiate the hoax from fraud or misconduct, as a hoax may be designed and perpetrated not for the benefit of the hoaxer but to confound others in their dream worlds. In that unique specimens may be implicated in a hoax, as the Piltdown Man, confirmation may not be feasible without examination of each artifact. The hoax intended to benefit the perpetrator, of course, is fraud.

The complexity in defining these terms and in determining which may apply in specific cases remains one of the great problems of biomedical science. Moreover, fraud is of two sorts, admitted or confessed and suspected but unconfessed. Then there are those unreproducible papers that are of doubtful validity, such as the scotophobin case and the memory of water at infinite dilution that are dubious science discounted over suspicions of error but not necessarily of fraud.

One sees from these few cases, there are oodles more, a number of characteristics. Fraud by established scientists is not so much the case as is fraud from those with ambition to become established, to become famous, to bask in fame for the nonce. Not that senior scientists may not be involved in fraud; however, their maturity ordinarily should acquaint them with the likelihood of rapid discovery, with the loss of any honors.

The student and junior investigator in a hurry for success sees in some odd way momentary fame through fraud as desirable. Several features are common to these episodes; the junior fraud is usually a member of a busy senior scientist's laboratory, uses the opportunity to fake results, and becomes instantly famous. The fraud perpetrator is invited to Gordon Research Conferences, to NIH *ad hoc* specialist meetings, and to many a laboratory for seminars in advance of journal publication, where they are lionized for their remarkable discoveries. Peer and editorial review is too slow, so the process is subverted. Abstracts may appear. Invitations to present the work before national and international symposia before publication may follow. Print and broadcast media provide publicity far beyond the ordinary.

Some of these same sequences may occur where fraud is absolutely absent. I have seen junior scientists lionized at Gordon Research Conferences over their work with established, honored senior scientists, work in press or just published. These lionized juniors may go on to their own independent careers with or without fizzle.

Fraud may occur where high stakes are perceived, by the perpetrator and by his mentor hungry for greater fame, genuine success, and more grant funds. The senior party is ofttimes a well known scientist or medical practitioner, someone in demand for all sorts of extramural activities that take the senior away from the laboratory and leave matters in the hands of less experienced junior associates. Senior investigators who travel a lot, attend many international conferences abroad, with large research laboratories, with heavy administration and funding duties, inadequate supervisory skills, and with attention to their own career successes but not to scholarly pursuits create easy opportunities for misconduct by junior associates eager to build their own meteoric careers.

The regular presence of the senior independent investigator in the laboratory makes faking evidence almost impossible. On the spot supervision, review of samples as produced, data as collected, and review of laboratory records as generated reduces opportunity for misconduct and fraud. As a reputation for close control ("trust, but verify") becomes known, those who would contemplate fraud tend to drift elsewhere.

It appears that many established scientists have had experience with fraud. Efraim Racker, victim of the Mark Spector fraud, noted several cases of fraud in the laboratories of Carl Cori, Washington University, St. Louis,MO, David Green, University of Wisconsin, Madison,WI, Melvin Simpson, Yale University, New Haven,CT, and Fritz Lipmann, Rockefeller University, New York City,NY. The miscreants were not named by Racker, but all four seniors retracted the affected work.

Self-discovered Fraud. Self-discovered error resulting in retraction of published works is now complicated by similar retractions of papers derived from fraudulent work. Both confessed and unconfessed fraud may be involved, as already outlined (Chapter 2). Retractions may follow upon failure to confirm results in the laboratory of origin or by others, in which case uncertainty of status follows. It is not always certain whether honest or foolish error be the case or whether unknown or unconfessed fraud be implicated. Moreover, in the current rage for litigation, honest appraisals of troubled work by involved scientists may not be possible for fears induced by threats of legal actions.[40]

Besides the cases previously described, self-discovery and confession of fraud and withdrawal of papers is exemplified in the recent case of Francis S. Collins, director of the National Genome Project, NIH National Center for Genome Research, who retracted in 1996 five papers dealing with the genetic alteration of nonexistent white cell lines overexpressing the gene CBFß leading to acute leukemia. A junior colleague Amitov Hajra admitted falsification of data.[41]

Far more devastating are frauds that are unconfessed, frauds that must be recognized from suspicion and failure to confirm questionable work, frauds that take much time and effort for resolution. Career through several important biomedical and clinical cases confirms the extent to which confessed and unconfessed fraud has blossomed. Clinical fraud involving patients and in medico-legal matters is of utmost importance in that innocent parties may be harmed.

Clinical Fraud Cases. Perhaps the most notorious case of recent biomedical fraud is that of surgeon William T.

Summerlin seeking to demonstrate cross-strain skin grafting in mice at the Sloan Kettering Institute. However, he was observed marking black spots on a white mouse with a felt tip pen by a colleague in an elevator and was thus exposed. He confessed to fabricating evidence, was considered to be mentally ill, and was given a medical leave.[42] We have another interest in this affair, as Summerlin had been a surgery resident at UTMB before going to Sloan Kettering. This aspect is not publicized locally.[43]

Physician John R. Darsee reported fraudulent work aimed at prevention of heart attacks, using dog models. His work at Harvard University Medical School from 1981 led to five papers in fifteen months with several coauthors. He also had published ten papers and 45 abstracts while at Emory University in 1974-1979, published work for which no records exist and with three nonexistent scientist coauthors acknowledged.

Darsee's fraud was initially discovered in May 1981 but was not reported to the NIH until 1983 when Darsee's fraudulent data had ruined a costly study involving others.[44] Harvard was asked to return $122,371 to the NIH. Stewart and Feder at the NIH concluded that Darsee's papers had so many problems that his 47 coauthors should have discovered them; coauthors had "not been candid" in reporting their work.[45] Publication of the account was delayed by worry over legal matters until 1987.[46] Noted cardiologist Eugene Braunwald, head of the Harvard laboratory involved, was accused of inadequate supervision because of his other activities; he took exception.[47] Darsee blamed the high pressure at Harvard and asked for forgiveness for whatever he did wrong, without admitting misconduct.

Charles J. Glueck, Professor of Internal Medicine, University of Cincinnati, was found guilty of serious scientific misconduct for misreporting growth of children on low cholesterol diets, censured by NIH in 1986, and barred in 1987 for two years from NIH funding. His paper on the matter was adversely criticized as not being a prospective study and as having internally inconsistent data. He resigned from the university faculty February 1987.[48]

Clinician Mark J. Straus, New York Medical College, Valhalla,NY, was charged with falsification of clinical oncology data. He had a three-year $1 million NIH grant but was denied further NIH funds in 1982. His published work apparently had faulty refereeing. He claimed he had no part in the falsified work but was framed by co-workers.[49]

The Burt case and that of Stephen E. Breuning next show that fraud extends to the psychologist as well as to the physician. British psychologist Sir Cyril Burt studied the IQ of twins reared apart; his data are alleged to be faked. He was also charged with slanting other authors' papers submitted to his journal to suit his notions and of misquoting adversaries repeatedly in debates. He is excused on the basis of his age and psychological decline.[50]

The prominent case of psychologist Stephen E. Breuning, University of Pittsburgh, who falsified results of testing of drugs on retarded children, is further instructive. Between 1979-1984 Breuning produced one-third of the psycho-pharmaceutical literature about mentally retarded persons. His work was influential in guiding medical treatments of retarded.

In this case, like the Poisson case, it was the head of the large National Institute of Mental Health (NIMH) project Robert L. Sprague, University of Illinois, who first questioned Breuning's data. After exposure, citations to Breuning's works declined sharply. He was charged with lying to NIMH by submitting fraudulent results in a grant application and of obstruction of NIMH investigations. Breuning admitted to "some mistakes", resigned from the University of Pittsburgh, and was later convicted of criminal fraud.[51]

Biomedical Fraud Cases. Many other cases are known. The following brief sampling of misconduct cases, not meant to be definitive, outlines the nature of prominent examples and more interestingly the nature of any punishment following proof of misconduct: (1) Obstetrician Harvey M. Levin, convicted of providing false data on a painkiller in clinical trials in Philadelphia,PA, in 1983.[52] (2) Delbert Lacefield, fabrication of blood test data on Federal Aviation Administration blood tests on people involved in train and airline crashes in 1987.[53] (3) C. Sibley developed a technique of DNA hybridizatiion, resolved phylogenetic relations between birds and primates, and was elected to the National Academy of Sciences. His manipulations of data discovered by others in 1988 were admitted by him in 1990, that there were data manipulations, that he used control data from one experiment in other experiments, moved correlated points on a scatter plot into regression lines describing the work. No retractions, no inquiries.[54] (4) Physician John Long, with a seven-year grant, faked a Hodgkin Disease cell line at Massachusetts General Hospital, Boston,MA. (5) John Spengler at Harvard Public Health suppressed adverse statistics data in air pollution study in 1989. (6) J. L. Ninnemann, University of Utah and University of California at San Diego, submitted falsified data of burn injury effects on the immune system in a $1.2 million NIH grant application. His assistant sued for three times the damages under the False Claims Act. Notably, Ninnemann had been at the Sloan Kettering Institute during the Summerlin investigations and had been unable to reproduce Summerlin's work.[55] (7) J. W. Gordon et al. claimed to have created an Alzheimer's disease condition in mice but retracted the report.[56] (8) Radiologist Robert A. Slutsky made gifts of coauthorship to others on fraudulent papers. He had one paper with faked data every ten days, University of California, San Diego, 1983-1984.[57]

Publication of very large numbers of articles in a short time provides suspicion of misconduct, plagiarism, or fraud. Virgil Percec, Case Western Reserve University, published 56 papers in 1991. Russian chemist Yury Struchkov published 948 papers in 1980s and 83 in 1991, got the Ig

Nobel Prize 1992.[58] However, such generalization is not reliable, as many productive scientists publish many articles per year. Nobel Prize laureate Derek H. R. Barton, recently deceased, published 39 articles in 1995-1996, all without blemish.[59]

Other Fraud Cases. Suspect fraud cases are neither new nor confined to biomedical and clinical research. Consider the *Gedanken* case of M. Lazzarini who reported in 1901 a value of π = 3.1415929... (true value 3.1415926...) calculated from an odd method of throwing a needle onto a grid and counting the number of times the needle lay on a grid line. Lazzarini reported dropping his needle 3,408 times. It was concluded that the report was fraud, that the experiment had not been conducted.[60]

An example in chemistry, the reaction of triphenylmethyl radical with nitrobenzene, was reported to involve radical abstraction of oxygen from the nitro group. However, the reaction could not be repeated, neither by G. S. Hammond nor by others.[61] The case is a perfect example of fraud by postdoctoral fellow A. Ravve but also of overenthusiasm and wishful thinking by Hammond.

More recent odd examples include those of Armando Garsd, Harvard University, dismissed after questions of statistical validity of EPA air pollution work and of Viswa Jit Gupta, University of Punjab 1969-1989, who provided expert paleontologists with fossils from strange places and then coauthored papers with them. His ammonite fossils claimed to have been found in the Himalaya Mountains were probably from Morocco. A critic remarked "I am convinced we are not seeing the results of honest mistakes".[62]

Among other common difficulties is the matter of senior faculty appropriating work of their students and junior colleagues without proper attribution or agreements. A case at Cornell University, Ithaca,NY, pitted then graduate student Antonia Demas against David A. Levitsky. Cornell University made an inadequate investigation that cleared

Levitsky, but Demas and members of her Ph.D. committee were dissatisfied with Cornell University's handling of the case. Levitsky who later became a member of her committee was charged with misconduct in his use of her thesis data without attributions, for his own purposes.[63]

The perpetrators of these frauds must have very strange thought processes that they regard momentary advantages over eventual discovery. Most science frauds are discovered one way or another as others attempt to replicate claims or to build on them to other results.

The misuse of research funds may be simple misconduct, where unnecessary or frivolous expenditures are made or inadequate supervision be the case. However, other misuse is criminal in nature and thus not one merely of science misconduct. Nepotism and hiring of paramours come under this venue, as does embezzlement, and fraudulent use of funds in support of a grand life style.

In that financial fraud is not science fraud, only a very limited treatment of the topic is attempted here. Examples of financial fraud range the whole gamut, from token adjustments to criminal prosecutions. The notorious case of the 1994 Office of Naval Research investigation of Leland Stanford University administrators who used research indirect costs to support their yacht resulted only in their having to repay $1.2 million. It was deemed that there was no fraud![64] Be kind to Stanford University.

More severe treatment is accorded some. In the strange misuse by Wasim Siddiqui, University of Hawaii 1984-1988, of perhaps $100 million in attempt to develop a vaccine against the *Plasmodium falciparum* malaria parasite, a goal for which there was little hope of success, the money was spent, but there is no vaccine. Siddiqui was charged with fiscal mismanagement and incompetence and indicted 1989 on charges of theft, criminal solicitation, and conspiracy.[65]

The lure of generous NIH funding for AIDS has now also attracted those who embezzle. Lionel Resnick, Mt. Sinai Medical Center, Miami Beach,FL, is charged with embezzlement of $570,000 for work conducted at Mt. Sinai but billed by his own company Vironc Inc. 1989-1994. Pathologist James R. Allen, University of Wisconsin, took ski trips using NIH grant funds, for which there resulted a criminal conviction 1980.[66]

5. Ethics

By ethics one means values, the word deriving from the Greek ethos (εθος). Besides the obvious ethics issues of cheating, lying, plagiarism, nepotism, and misuse of funds, we must now by fiat accept sex/race harassment, unobserved affirmative action, civil rights and animal rights, and failure to adhere to government regulations as serious ethics offenses. We are faced with deliberate, designed use of words to mislead, use of acronyms, slogans, cliches, loaded words, euphemisms, and PC, ignoring published contributions of others, prominent display of name and self-citation, courtesy authorships extended and accepted, citation only of important people and name dropping, publication by press conference and abstracts only, stone-walling whistleblowers, and much more such that there are now demands for creation and imposition of ethics standards in research organizations.

Journal Publication Policies. Among other science societies the American Chemical Society has created a series of ethics guidelines related to journal publications.[67] Ethics for authors, editors, referees, and those publishing outside normal channels have been devised. For Authors: (1) Present an accurate account of your work. (2) Be concise in writing. (3) Include sufficient detail to enable peers to repeat your work. Provide sources of materials and character of procedures used. (4) Make your manuscript useful to those not expert in the subject. (5) Cite prior publications sufficient to place the manuscript in a proper intellectual context. Citation of pertinent review articles suffices. (6) Identify unusual hazards associated with the work. (7) Avoid publication of fragmented work. (8) Provide the editor with copies of related manuscripts in press or under editorial consideration. (9) Do not submit the same material to more than one journal at a time. (10) Do not disclose information obtained privately save with permission of those providing the information. (11) Do not put personal praise nor adverse personal criticism of others in submitted manuscripts. (12) Authors may request certain persons not be used as referees. (13) Coauthors of a paper should be those who have made a significant contribution. All living coauthors must approve the manuscript submitted for publication.

For Editors: (1) Give unbiased consideration to manuscripts submitted. (2) Review manuscripts "with all reasonable speed". (3) Seek advice regarding all submitted manuscripts. (4) Do not disclose information in a manuscript other than to the referees. (5) Respect the intellectual independence of authors. (6) Manuscripts of an editor submitted to his own journal must be reviewed by another editor. (7) Do not use information in a manuscript prior to publication save by the authors' permission. Editors should recuse themselves where a submitted manuscript is closely related to publications of the editor. (8) Upon learning that a paper published in his journal be erroneous the editor should arrange for publication of a statement of error and a correction from the authors.

For Referees: (1) Scientists have an obligation to serve as referees of submitted manuscripts. (2) Judge the quality of a manuscript with respect to high scientific and literary standards. (3) Referee objectivity requires that referees recuse themselves where there be conflict of interest, or provide a signed statement making explicit the matters in conflict. (4) Provide specific remarks of criticism. Unsupported claims are of little value. (5) Note any lapse in citation of relevant literature. Any similarity between a submitted manuscript and another submitted concurrently to another journal should be noted. (6) Submit critiques within the time limits suggested by the editors. (7) Treat submitted manuscripts as confidential items and do not disclose material to others. Others aiding in manuscript review should be disclosed to the editors. (8) Do not use unpublished work of authors save with the authors' permission.

For Those Publishing Outside Normal Channels: (1) Scientists publishing in the popular literature have the same obligations to be accurate as were they publishing in a science journal. (2) Less exact terminology may be necessary for communication with lay public, but public disclosures should be as accurate as possible consistent with effective communication. (3) Proclamation of discovery should not be made save experimental, theoretical, or other support is sufficiently strong to warrant publication in science journals. Public announcements should be followed shortly by publication in science journals.

Also encompassed here is the failure of authors to disclose financial interests in work submitted for publication. Along with misconduct, potential conflict of interest has become a major ethics issue in biomedical research where so many have formed their own businesses to exploit their science results.

Medical Ethics Courses. All our students are now exposed to ethics issues in required course work. Medical students take a separate required medical ethics course, and as required by the NIH all first-year graduate students (including nursing and allied health science students) and all predoctoral candidates and postdoctoral fellows supported on NIH funds undergo a special course in ethics "Ethics of Scientific Research". At UTMB the required course created in 1991 is conducted over a four-day period by our Institute of Medical Humanities for one credit-hour.

A NOVA television documentary "Do Scientists Cheat?" is shown at first, after which science faculty participate in such presentations as: "Must Good Science be Ethical Science?", "Why the Regulation of Research", "The Western Tradition of Science", "On Being a Scientist", "Responsible Conduct", "The Sloan-Kettering Affair", "The Harvard Fraud Case", "When Misconduct Occurs: Whistleblowing, Due Process", "Science and Society: Mutual

Accommodations", "Research on Human Subjects", "Ethical Dilemmas in Scientific Authorship", and "Conflict Resolution" and the like.

In our criticism course we addressed some of these topics depending on timing and opportunity but did not cover those topics emphasized in the NIH required ethics course.[68]

Ethics Challenges. Despite best efforts, serious ethics issues have arisen at UTMB, four examples being cited here for balance. Our medical school has the unique honor of having witnessed an ethics matter of great heuristic value. Twin brother medical students Forrest and Wendell Wall, sons of a surgeon, were discovered to have submitted the same term paper, perhaps a purchased paper, in the required medical ethics course. Though the twins admitted cheating, President Thomas N. James assigned relatively mild corrections. However, the twins were not allowed to return to classes in the next term, so they sued claiming their rights to due process were violated. A state judge agreed and forced their readmission, to the consternation of UTMB faculty and students.[69] The school's honor code was destroyed in the process; no longer must a medical student swear they will not cheat. Lesson: If the rules are too stringent, drop them.

The same problem may have developed in other places where honor once ruled. We hear that the U.S. military academy might no longer enforce the oath "I do not lie, cheat, or steal, and I do not tolerate those who do". Could this be so!

There always arises the question what would you do if you witnessed misdeeds of others, including those of administration, faculty, students, and staff. Some years ago a Dr. Fell appeared among UTMB faculty, he purportedly having a doctoral degree, which was not the case. By story, several faculty aware of the case informed administration, and some correction occurred. On a more recent occasion a faculty member used research funds to acquire a fish-finder for his boat; a disgruntled lab assistant informed on him, and he left the University.

Yet more egregious but with a different result, an ambitious well-funded faculty member became involved with much publicized clinical trials of a substance he had prepared. The trial involved children, but the institutional committee required to review proposed clinical trials had not been advised nor had approval been granted. All that resulted in the way of correction was a file on the matter kept by an assistant dean.

We see many such ethics challenges among the numerous papers and books now being published that address these issues.[70] Test yourself. In the following instances, what would you do: (1) A potential competitor requests cultures of your published mutant wonder cells. Do you send the parent strain or the mutant? (2) You know that a technician in the next lab accidently spilled ^{131}I-labeled material but hid most of the contaminated materials in another lab before calling the radioisotope safety office. Do you report this? (3) A graduate student is overly solicitous of their and your faculty mentor such that you suspect or recognize unprofessional favor developing. (4) You as the teacher are asked to grade your quiz differently so that a favored racial minority student gets a better grade. (5) A student's term grade you turned in to the office was altered by the chairman or dean without your knowledge or approval. (6) You observe a lab-mate taking shortcuts in established procedures and altering protocols to suit their schedules, abilities, or interests. (7) Your lab work is used in a journal manuscript, but your name is not among authors nor do you get an acknowledgement. (8) A faculty member ostensibly travels abroad to a science meeting but actually does not go to the meeting and takes a vacation using travel funds provided by the company, university, or research funding. (9) A senior investigator asks or demands coauthorship of a manuscript for publication without having done anything on the project. (10) Your latest manuscript offends a local politically powerful person whose views are at odds with your own. What to do?

Indeed, what to do. Go with the flow? Accede to dishonor? As in all walks of life, one makes their way as best they can.

In summary, besides the age-old Golden Rule, very simple rules for conduct are formulated: Be honest; never manipulate data; be precise; be fair with regard to priority; be without bias with regard to data and ideas of your rival; do not make compromises in trying to solve a problem".[71]

CHAPTER 7. SOME OTHER MATTERS

De quoi s'agit-il?
(What's the Problem?)
- Marshall Ferdinand Foch, 1918

Over the course of each term of class work we delved into several recent science matters receiving media publicity and government sanction. At the emergence of each topic issues were raised of authenticity, of seriousness. Not every topic was discussed in each term, but major items were examined for contrast, how to tell the real from the contrived or fake. How possibly to deal with the problems associated with each item.

Among the topics are those that involve genuinely new ideas and advances in biomedical science, topics for which Nobel Prizes are awarded. Other topics include those of pure foolishness (Polywater), official dogma (disease of the month, cholesterol the villain), dubious science that still attracts adherents ("cold fusion"), offensive matters (PC, commercialization and conflict of interest), future danger (loss of tenure), knowledge of the literature, and government regulation of science (research funding, federal science police). Polywater and "cold fusion" have been discussed in Chapters 2 and 5, the horror of PC in Chapter 4. I discuss here the disease of the month, cholesterol as villain, conflicts of interest, tenure, knowledge of the literature, and government regulation of biomedical science.

1. Disease of the Month

We ofttimes emphasized the consequences of rapidly increased federal funding leading to bandwagon parades. Obviously many very valuable advances have been accomplished by generous federal funding, but with the money come regulations. Moreover, federal funding is influenced by social and political considerations. The Disease of the Month is actually the disease of the current political climate, the health disorder most favored by each President of the United States. Inasmuch as President Lyndon B. Johnson was afflicted with heart problems, so the National Institute of Heart and Lung was funded generously. President Richard M. Nixon's interest in cancer led to improved funding of the National Cancer Institute (NCI). President James ("Jimmy") E. Carter favored mental health. The recent emphasis on AIDS and HIV virus infection reflects President William ("Bill") J. Clinton's interest in those afflicted.

It is obvious that increased funding ultimately results in improved understanding of each health disorder, with eventual changes in medical treatments. The allocation of NIH research funds is instructive in this matter. The actual NIH funding in 1997 per death attributed to an identified health disorder was $2,000 for heart disease, $5,000 for cancer, but $70,000 for AIDS. The extraordinarily increased funding for AIDS in the face of relatively low numbers of persons afflicted clearly reflects the political influence. Moreover, we have World AIDS Day celebrated annually December 1. Do you recall a World Heart Day, a World Cancer Day?

Anent the AIDS problem, the question whether HIV virus infection be the necessary and sufficient event to cause the affliction has been raised. Current thought is that HIV infection causes AIDS. However, Peter H. Duesberg, University of California, Berkeley, states that HIV infection *does not* cause AIDS, that HIV is merely a marker for AIDS.[1,2] For his efforts Duesberg has been vilified. As with other studies in which homosexuality is involved, departure from orthodox thinking attracts demonization.[3] The matter presently appears to be settled in favor of the official version funded by the NIH but also dogmatically asserted in the Durban Declaration, the 13th International AIDS Conference of 2000.[4]

When heretical concepts conflict with official versions, the serious question must be cast: is either version correct? We know that both cannot be correct, but that both can be wrong, and that yet other understanding be required. Recall the related case of the controversial work of Stanley B. Prusiner, University of California, San Francisco, Nobel Prize laureate in physiology or medicine 1997, establishing the concept of the prion as protein sans nucleic acid as an infectious agent causing disease.[5]

The prion protein concept is implicated in slow virus diseases such as scrapie, kuru, Creutzfeldt-Jacob disease, and mad cow disease (bovine spongiform encephalopathy), but it has proven difficult for some to accept the concept. The prion has been termed the "so-called" prion and called "bizarre" in published articles. Nonetheless, NMR spectra have revealed recently that variant protein folding of recombinant prion protein fragments may give an infectious conformer,[6] thus providing strong support to the concept.

2. Cholesterol The Villain

In this section we examine effects of the official, government-supported crusade against cholesterol as cause of coronary heart disease. From time to time, official crusades have been mounted, for good or no, against target entities, most recently cholesterol, firearms, and tobacco products. We even have a "War on Drugs" against abused illicit drugs, patterned after the failed prohibition of beverage alcohol of long ago.

In these crusades the proscribed agent is *the* cause of unwanted results. The keyword alerting serious considerations on these issues is "cause". Causation of natural events is a complex matter that baffles the unwary but accords the advocate of a contrived notion means to gain their ends by confusing or avoiding definition of the word.

The anti-cholesterol crusade is a perfect example where

causation was asserted but where there is no convincing evidence of cause and effect, and alternative views were not considered. Crusaders cannot conceive that elevated blood plasma levels of cholesterol and cardiac disease have a common cause, that both are effects that may co-exist without one causing the other. The official government and food industry position has been that cholesterol is bad, be the villain, causes heart attacks, and should be outlawed like illicit drugs! We have seen a crusade against cholesterol mounted to reduce blood plasma levels of cholesterol by dieting, drugs, or other medical interventions.

The crusade started well over two decades ago, during which time a desired level of total cholesterol in blood was set, demonization of red meat, dairy products, and eggs occurred, and other issues were ignored. Advertisements proclaiming "Say NO to cholesterol", "Down with cholesterol", and the like were common. At one time, perhaps in jest, lawsuits were proposed in California to sue food manufacturers whose products contained cholesterol!

I once in all innocence at a nutrition meeting dinner made some remark about cholesterol not being the villain that the crusade required. The young woman nutritionist at the dinner table almost had a fit; she knew cholesterol *causes* heart trouble.

The idiocy is extreme in other cases. The 1994/1995 Calbiochem Co. chemical catalog lists purified cholesterol with the awesome, auspicious warning "WARNING! May be carcinogenic/teratogenic".[7] This catalog warning must come from the same attempt to shield manufacturers from wicked tort claims like that of the man who fell off his ladder and sued the ladder manufacturer successfully, on the grounds that they did not tell him of the danger of falling off a ladder. That cholesterol, like water and oxygen, is absolutely required for cellular life escapes the attention, or even interest, of the crusaders, as they cannot explain how an indispensable cell component is in itself so dangerous that government intervention must be had.

More recently the offensive crusade against cholesterol

has abated; federal government and processed food manufacturers are altering their stance, as they seem no longer convinced they are tackling the right villain. We now see regularly "bad" low density lipoprotein (LDL) cholesterol but also "good" high density lipoprotein (HDL) cholesterol.

There is an emerging awareness that cholesterol so intimately related in some yet obscure fashion with the etiology of human atherosclerosis is not the cause of heart disease. Rather, intravascular oxidation processes currently under intensive study seem to be implicated as a major factor, if not cause. The artery is, among other things, an oxygen pipeline, and oxidized blood components now receive energetic attention aimed at determining just how inspired oxygen oxidizes tissue and blood components and how these oxidations are implicated in ultimate cardiac problems. We hear of no federal government demand that our oxygen intake be restricted. Indeed, emphasis is now shifted by government, food manufacturers, and drug companies to antioxidants as a means to improved health.

3. Commercialization; Conflict of Interest

Today there is a threat to science progress perhaps greater than that of misconduct though not now recognized as such. The appeal of misconduct as vehicle for expanded government control of science takes priority over that of conflict of interest not yet considered to be misconduct.

In the halcyon days of yesteryear science faculty pursued teaching, research, and scholarly matters as best they could. There were esoteric discoveries and inventions apparently without commercial potential or business interest. Recall British Prime Minister Robert Peel's 1831 inquiry of Michael Faraday's dynamo "And what use is it?", and Faraday's response "I know not, but I wager that your government will eventually tax it".

Although some had arrangements with commercial enterprises for modest research support and some work was patented, business interests were not emphasized nor predominant. Now the intrusion of business interests and practices into academic pursuits has become quite common, particularly so in biomedical work currently in favor dealing with recombinant DNA, gene sequences, etc.

Conflict of interest comes in different forms. As previously noted, the funded researcher has allegiance to his source of funds, not necessarily to his own institution. This conflict of interest is tolerated by the institution as indirect costs come to administration for administration use not related to research. The conflict thus includes both researcher and his organization.

As faculty members move deeply into their business enterprises yet retain their academic appointments and emoluments, further conflicts of interest result. Publicly funded research results must be exploited for personal gain by patenting, by licensing to commercial organizations, or by founding one's own company. The lure of great personal wealth attracts many.

As these matters progress, the same problems of the business world descend upon the realm of experimental science. Without consideration for the financial gains of business success or of failures whose costs are borne by public funding there is the question of impact on intellectual processes and on science. There is the matter of secrecy attendant upon patent protection of commercially exploitable work, conducted at public expense but accorded the Wunderkind inventors as a matter of favor.

There was a day, still is, when pharmaceutical industry scientists would go to meetings, present their work, but decline to answer certain questions that went beyond what the patent lawyers had approved for public discussion. One became accustomed to this secrecy despite the distaste, but the same secrecy among academics trying their hand at immense wealth is now pervasive in biomedical efforts.

Universities set up special administrative offices to encourage and manage commercialization of research. For instance, the University of California, San Francisco, has an office of technology management that deals with relationships among faculty, administration, and commercial businesses. There is a material transfer agreement between the academic institution and the business in which all sorts of arrangements for sharing wealth are devised, also that place restraints on disclosure of information and which act to advertise the materials or services available from the business enterprise.

Such business practices lead to another item of concern, that of vulnerability to lawsuits. My own recent experience with such matters was with a federal government laboratory that provided me with a toxic marine peroxide for spectral studies, along with a formal release to sign. The release as much as said that if I ate the sample the government was not responsible. For decades individual investigators shared their precious steroids with me, as reference compounds, without concern for litigation. Perhaps no longer, given offices in every institution that regulate such matters.

Other problems now arise regarding results obtained with public funding. What are "data" and who "owns" them? At our University of Texas Medical Branch these two issues have been resolved by fiat of Assistant Dean of Medicine Walter J. Meyer. Data include written information, that stored electronically, chemicals, cultured cell lines, viruses, and other material items. Data are owned by the University; faculty originating data are merely custodians, not owners![8]

Definition and ownership remain uncertain for data acquired under other, nonfederal auspices, but extension of federal control to all research results however funded seems wanted. Several bureaucratic suggestions to control research have been advanced, including Congressional suggestion that the now defunct OSI conduct random audits of laboratory records, thereby to ensure somehow against misconduct.[9] Further, in the guise of assigning accountability to investigators funded by the federal government the Government Performance and Results Act of 1993 and the White House Office of Management and Budget seek to extend the Freedom of Information Act (FOIA) to cover data of funded researchers, including handwritten laboratory notes, field notes, and computer tapes! Thereby funded research generates data that are directly in the public domain, to be used by whomever for whatever purposes however far removed from science.[10] The funded researcher is now clearly to be a government ward or employee!

Use of the science and medical literature for personal and business advantage poses another conflict of interest. Should financial interests be disclosed in the publication of a process or procedure bound to benefit the authors. The expansion of biotechnology creating new drug candidates has increased need for disclosure of personal financial interests. Disclosure of consultancies, patent rights, stock and stock options, and expert testimony in lawsuits is now required by some medical journals but not by all.[11]

There is additionally now a new matter, the alteration of time-honored publication standards of the journal *Science* whereby a paper announcing the alleged full sequencing of the human genome by the Celera Genomics Co. is planned, but without the usually required full disclosure of the human genome data, thereby to protect the commercial interests.[12]

These conflicts of interest have been with us for decades in medical journal articles, where physician authors may greatly influence commercial matters. In the heyday of discovery of new antibiotics in the 1950s the acceptance for publication of papers from ethical pharmaceutical houses by a notorious journal dealing with antibiotics was dependent on purchase of vast quantities of reprints. Accelerated publication required still further purchases.

Despite disclosure provisions there still appear medical articles by persons variously paid by pharmaceutical companies for their expertise. Editorial oversight of a sex dysfunction article in *JAMA* requiring subsequent journal announcement was claimed.[13] However, there remains the question whether disclosure of financial interests, particularly of corporate or agency funding, unnecessarily biases the article. By critical scrutiny of the sort we advocate in our course the validity of a study should be addressed on its merits and not by funding sources, university of institution affiliation, or names of authors.

Additionally, ever so many faculty members have lucrative private consultation agreements with business as well as their own private businesses conducted on university premises and funded by public money. These arrangements necessarily create conflicts of interest. Researchers seeking their own personal success, fame, and fortune obviously must meet business needs before those of science and the public. Nondisclosure agreements, confidentiality agreements, patent applications, corporate permission to publish, censorship or withdrawal of science papers, gag orders, threat of lawsuit, and sudden dismissals characterize the situation.

Ever so many biology advances have been incorporated into a faculty member's private company for advantage, for commercial exploitation. Find a gene, discover a treatment, build your own Microsoft. Why not get rich too.

At UTMB there are at least two faculty in our HBCG department who retain their appointments but who have established their own molecular companies to exploit their expertise gained while faculty members. These arrangements are approved, indeed encouraged, by administration. Massive influx of business funding for research conducted in the universities is much sought. Lucrative arrangements for funding in exchange for rights to anything the commercial ventures want is now a reality.

Faculty need not teach or provide other services, and the university hires others for such efforts. Usually advertisements for faculty are for the junior untenured ranks but with the proviso that the candidate must have an impressive history of research accomplishment, an exemplary record of external research funding, and finally, a strong interest and skill in teaching. Teaching is always mentioned last in order, as new faculty are more expected to fund administration through research grant overhead.

Some recruitment advertisements for faculty describe requirements such that no one may qualify. Only senior persons are of interest, they being those already with big research funds that can be transferred. Entry level faculty appointments are fewer, as one takes a chance on new hires who may not be successful in arranging funding rapidly. Where tenure still exists, the probationary period is sometimes extended considerably, thus keeping faculty untenured far too long and making them vulnerable to whim and fancy, keeping them as indentured but hopeful servants of administration.

The issue of conflict of interest is not confined to molecular biology, as the commercialization of discoveries and developments has long been with us in other fields as well. Three examples in chemistry reveal how far the problem can go.

NMR Spectroscopy Diagnosis of Cancer. The use of NMR spectroscopy for cancer detection has been a hope for decades. From 1971 Raymond Damadian used relaxation times of water protons in rat tumor tissues as measure, relaxation times being longer in malignant tumors (Walker sarcoma, Novikoff hepatoma), lowest in normal tissues, with benign tumor relaxation times being intermediate. In concept, tumor water experienced greater organization in cells, leading to longer relaxation times.[14] Damadian extended his claims to excised human tumor tissues. However, the ranges of ^1H NMR signals for malignant, other abnormal, and normal tissues overlapped such that the test was unreliable. Neither results nor claims were confirmed by others.[15] When queried about original data in his human tumor studies, Damadian responded "Your group seeks to cheat me of credit for my discovery. You expect me to provide the data as well?". He seemed obsessed with who got credit.

Credit is always of great importance. One may be surprised at what can be accomplished if there is no concern for whomever gets the credit. Others hog credit mercilessly. A key example is the discovery of streptomycin by Albert Schatz working in Selman Waksman's laboratory that accorded Waksman the 1952 Nobel Prize. Schatz was ignored for the honor.[16]

Damadian later reported use of ^{31}P NMR relaxation times as measure of malignancy in excised tissues, publishing the same data twice.[17] From these studies he devised a concept of a single NMR signal diagnostic of malignancy, with which a cure might be had by irradiating *in vivo* the patient at the frequency of the unique signal, thereby heating the tumor and killing it. However, such approach could not heat tissue so as to kill.[18] In other work there was claim that malignant tissues had lower, not higher, levels of ATP.[19]

Despite the thorough effectiveness of standard pathology examinations of excised tissues for malignancy and the lack of acceptance of his methods and the ultimate recognition of their unreliability, Damadian formed his own company FONAR Corporation to exploit matters. He projected the notion that NMR examination of surgical tissue samples would shortly be the means by which malignancy would be diagnosed, that there would be a NMR spectrometer in every physician's clinic for use in diagnosis.[20]

A more recent attempt to use NMR data for cancer diagnosis teaches us further about such matters. Whereas Damadian used NMR data from excised surgical tissue samples for diagnosis of malignancy, Eric T. Fossel proposed use of human blood plasma samples. The empirical measurement of line widths of human blood plasma lipoprotein methylene (CH_2) and methyl (CH_3) proton signals was advanced as means for diagnosis of cancer.[21,22] Within a month the report was public news.[23] Line widths of the proton signals in water-suppressed NMR spectra of human plasma components, generally of lipids, were 39.5 ± 1.6 Hz for normal plasma samples, 29.9 ± 2.5 Hz for samples from patients with diagnosed malignancies.

The report set off a furor of efforts to examine the

procedure. A Blood Plasma NMR Group was organized at the 1987 Experimental NMR Conference at Asilomar,CA, under the auspices of NCI. Information interchanges occurred; by May 1,1987, a list of 77 laboratories working on the problem, of which ours was one, had been compiled. Meetings in August 1987 of the Society of Magnetic Resonance in Medicine and at NCI September 28,1987, dealt with the matter. A standard experimental test protocol was devised.

Confirmations but also mixed results of overlapping data appeared shortly,[24,25] with major admonitions appearing by early 1988.[26] Factors not considered by Fossel *et al.* greatly influenced values. Fossel *et al.* used thawed plasma that had been frozen; no dietary factors were considered. Other factors not considered by Fossel *et al.* were directly shown to be important. Diet was crucial; postprandial lipemia caused line width narrowing, thus false positives, as did high plasma triglyceride levels and high very low density lipoprotein (VLDL) levels. Other factors included treated versus untreated cancer patients, sex, age, pregnancy, use of plasma but not serum, and need for high field NMR spectrometers at 360 or 400 MHz.

By 1990 the test was recognized as unreliable.[27] As is common in such cases, counter arguments emerged.[28] Our own work on this project involving many plasma sample analyses yielding some support was abandoned for want of clear reliability.[29]

Of fundamental importance to the suggested diagnosis of cancer by these methods is the question of ethics, of how to deal with false positives! Should a patient be told. Many a physician with whom I discussed this matter merely asked why would you want to know; what could you do about it.

In both these misguided cases only the time and expense of the efforts were the loss, as either proposal had immense potential for utility. However, there really was no loss, as these events serve well to establish the capacity of biomedical science to test matters, to determine validity or lack thereof. If you do not get something, you learn something.

However, as with ever so many technical advances, there are those who seek stipendiary gain. Although high field, superconducting NMR instruments were suitable to conduct the Fossel test, one entrepreneur George Sarantakos, Medic Matic International of Garland,TX, advertised for sale or lease a low field iron core fixed magnet NMR instrument for the task. Recall Damadian had also offered an iron core fixed magnet for diagnosis. Whether such magnets be suitable or such measurements be meaningful, no matter.

Cold Fusion Cells. In like manner, the flap over electrochemical cold fusion continues. Innovator James A. Patterson, Clean Energy Technologies Inc., Dallas,TX, now offers his cold fusion Patterson Power Cell kit for sale. Do your own experiments.[30] Related electrostatic confinement devices may also be marketed, not as "cold fusion" power sources but for creation of a neutron flux.[31]

Plagiarism and Commercial Exploitation. Plagiarism, commercialization/conflict of interest, and political influence at the highest level of government are implicated in this case. Hector DeLuca well known for his work with Vitamin D (calciferol), University of Wisconsin, Madison,WI, became embroiled in legal matters over alleged misuse of a chemical synthesis manuscript he had received for refereeing. The case is also a matter of conflict of interest as the Wisconsin Alumni Association Foundation, University of Wisconsin, Madison,WI, obtained a patent for the Vitamin D derivative 1α-hydroxycholecalciferol of potential medical use. The University had a long interest in royalties from early Vitamin D research, and prospects were good that much more was in the offing, given the patent. However, the synthesis described in the patent appears not to be the one actually used, and a lawsuit by no less than Nobel Prize laureate Derek H. R. Barton and Robert Hesse sought invalidation of the DeLuca patent.

It appears that DeLuca used a synthesis similar to one in a manuscript of Barton and Hesse that Deluca had received for review. Misuse of the Barton-Hesse manuscript was concluded in an initial formal investigations of 1988, at a time when Donna Shalala had become chancellor of the University. Other charges of misconduct were also made. However, a second investigation panel found DeLuca innocent of all charges.

Subsequent OSI investigation found apparent additions to laboratory records not made originally by the technician who proposed the original synthesis scheme at question. In 1990 Shalala stopped cooperation with OSI and admonished OSI to cease investigation. When Shalala subsequently became Secretary of Health and Human Services, thus controlling NIH fraud investigations, two weeks after her confirmation on April 9,1993, she ordered NIH investigators Walter W. Stewart and Ned Feder to discontinue their work on fraud.

The fraud investigations of Stewart and Feder were deemed beyond the mission of NIH scientists. They were reassigned to the National Institute of Diabetes, Digestive & Kidney Diseases, but the two refused reassignment. They were locked out of their offices, their data taken from them, their files impounded. They were further ordered not to continue their fraud investigations or use NIH supplies, stationary, or other resources. They were forbidden to discuss their government work of the prior ten years![32] Stewart began a hunger strike May 1993!

4. Tenure

Across the nation there is the call for revision of tenure for university faculty.[33] Somehow the tenure system has failed. Despite the great advances made under it, tenure must be drastically modified if not jettisoned. Where we once could afford the "unproductive" scholar indefinitely for the good of scholarship, the demand for instant success, immediate exploitation of breakthroughs, accountability, fund raising, and business methods of cost reduction now require destruction of the system.

Of course, the problem is that far too many faculty have been tenured for nonacademic reasons, for favoritism and political expediency, to build fund raising institutes within the university but outside the usual teaching and service duties. The increased costs of tenured faculty once their fundability days decline, superfluous faculty with no teaching duties, drives administration to desperate measures of cost containment. The bloat of administration must be preserved, as their work is important, but faculty that do not pay their way are to be untenured as directly as can be arranged.

The call to end tenure is all the more clarion in medical schools where physicians must fund administration through patient care fees and science faculty must do so through generous federal research grants. The insistence on fund raising also comes from external factors; increased competition between managed health care organizations and medical school hospitals leads to budget uncertainties. Although expansive administration must be preserved, unproductive "deadwood" faculty are expendable. Two business methods are advanced: cut salaries of those faculty who merely teach and do not fund administration and remove the protection of tenure so that unproductive faculty may be dismissed for not paying their way. The American Association of University Professors Subcommittee on Medical Schools reported to an Association of American Medical College conference on tenure in 1996 that the 1940 American Association of University Professors standards regarding tenure may not apply in medical schools.[34]

We see headlines "Tenure Has Outlived Its Usefulness".[35] Of course, we ask usefulness for what, to whom? We must have a faculty that serves administration. After all, we all want deadwood out and funded researchers in! Like other PC positions, this is a knee-jerk issue that lets you know quickly where one stands. Never define "deadwood"; use "unproductive" or "nonproductive" instead, however also undefined. The code words "accountability" and "flexibility" are found in anti-tenure pronouncements. These items mimic other PC modes of attack on other issues.

And who is to say who be unproductive: administrators seeking to reduce costs to administration. Administrators never trim administration, do they?

There now are two aspects of tenure, that of acquisition and that of retention. Ever so many problems of faculty who do not meet professional or political standards for tenure appointment have long been known and discussed. However, now along with the usual matters of professional accomplishments is the question whether faculty who do not obtain patents on commerciallizable products should be tenured!

With secret specific standards for tenure promotion both grant and denial of tenure is controlled by administration. Even in lawsuits brought by rejected candidates specific standards for tenure promotion are not disclosed lest the standards become known and met by candidates not otherwise suitable. We had such a case at UTMB some time ago. In a legal proceeding brought by an untenured female assistant professor against UTMB for sex discrimination in tenure promotion the then Dean of Medicine would not describe in the witness chair, under oath, what standards were used for tenure decisions. Necessarily a poor showing, and the jury found for the complainant. Also, other personal factors were implicated in the lawsuit, the point here being that arbitrary and personal arrangements may be introduced into the tenure system, this not ever being a secret.

The issue of publishing one's way to tenure is not always successful, as other sensitive matters become paramount. Recall that God never got tenure because: (1) He had only one publication, in Hebrew, with no references, not published in a refereed journal. (2) There are doubts about who wrote it. (3) What has he done since. (4) There is no replication of results. (5) He did not apply to the Ethics Board for permission to use human subjects, covered up mistakes by drowning subjects, and deleted from sample those subjects who did not behave as predicted. (6) He rarely came to class, merely told students to read the book, had infrequent office hours, usually held on a mountain top, and had his son teach certain classes. (7) He had only ten requirements; most students failed them; he expelled the first two students for learning too much.

In Texas termination of tenure has always been proper for just cause and for financial exigency, but by ending certain programs in which a decline in enrollment is claimed tenured individuals and departments also have been dismissed by inventive administrators. Dismissal of tenured faculty for filing grievances against the administration also occurs. A notorious case involving biochemist Wendell W. Leavitt at Texas Tech University Health Science Center, Lubbock,TX, found him dismissed, locked out of his laboratory, and electric power to his laboratory cut, ruining some experiments.[36]

Growing complaints from nonscholars that tenured faculty have a lifetime sinecure from which they cannot be ejected, that deadwood grows once tenure is awarded, that tenured professors simply are not worth it, have stimulated legislators to enter the fray. Under threat from Bill Ratliff, Texas Senate Education Committee, that either the University of Texas revise tenure regulations or the legislature would do it, the then Chancellor William H. Cunningham devised new conditions for retention of tenure, conditions adopted by the Board of Reagents and in place September 1997. Tenured faculty shall be subject to rigorous review of tenure every five years, with each component institution making its own standards. "Below standard peer and student evaluations for two consecutive years would provide cause for tenure revocation or dismissal"![37] But there is always question of who is peer to whom. 'Tis said the tenure policy exempts tenured faculty acting as administrators from such arrangements. Administrators do not criticize one another adversely nor do they teach; thus neither peer nor student evaluations are possible.

Regarding pending legislation, the intimidation reminiscent of the McCarthy anti-communist era led State

Senator Steve Ogden to ask: "What problem is this bill trying to solve?", "How do we keep this bill from having the very real, unintended consequence of being used as a political weapon?".[38] Legislature intrusion into the University of Texas is not new; faculty and students were required by legislative action to sign anti-communist oaths in the late 1940s period. This occurred because a political science or economics student had provoked legislators by making irritating comments to them while the legislature was is session.

I have a long list of the ever so many subversive organizations that one must not have joined be they employed as faculty by the State of Texas. Besides the Communist Party, membership in such organizations as Sons of the Rising Sun, Garibaldi Society, and Dante Alghieri Society (for certain years only) *inter alia* was proscribed.

Thus, the problem of maintenance of tenure is now a reality in Texas, where both the University of Texas System and the Texas A & M System have instituted tenure review policies. All manner of new arrangements to reduce costs of faculty are planned. Annual progress reviews, plans for the next year, and the now implemented five-year recertification of tenure promise great things. Nine-month appointments replacing the usual twelve-month appointments in medical schools, with guarantee of only some of the annual salary, is in process, as also is suggestion that all faculty take a 50% salary cut "lessening the pressure on junior faculty as well as keeping senior faculty on the grant track". Salary reductions can be recovered by your federal research grants and contracts, physicians by increased patient care billings. You pay yourself to raise funds for the administration.

Other schemes include longer probationary periods of up to eleven years for tenure consideration, switching regular academic appointments to non-tenure tracks and to lower-level instructorship ranks, fixed term contracts of tenure (renewable), and part-time faculty (avoiding tenure, fringe benefits). There is even talk of hiring substandard teaching personnel, those with education degrees who know how to teach even if they do not know their subject, what to teach.

The latest ploy devised at UTMB to end tenure is the Faculty Compensation Plan that reduces the salary of tenured faculty to the extent that they will resign their posts. The Plan is to guarantee 80% of the most recent American Association of Medical Colleges standard for the 50th percentile of comparable faculty. Those wishing to regain their full salary may do so through the incentive plan, get a research grant. The "at risk" portion of salary may be set arbitrarily by administrators; tenured faculty retain their academic appointment but not their salary.

As with acquisition of tenure, so retention of tenure in these circumstances (but not salary reductions) will probably be no problem, as once the standards become known faculty will devise means of meeting them. One merely manufactures the necessary credits, however, at the expense of scholarship. As grantsmanship has displaced scholarship, so survival skills will lead to whatever results are required. Administration will not get the cost advantage thought.

The protective mode at expense of scholarship occurs in other matters too. Twenty-five years ago after competition with UT Southwestern Medical College, Dallas,TX, for State budget increases regularly brought up the matter of how many more of their medical students passed the National Boards on first taking than did ours, our HBCG department went over to teaching to the National Board test that each medical student had to pass. Grades improved. Depth of ancillary materials taught declined. Lesson learned.

There is question whether faculty should be held responsible for budget shortfalls for which faculty are not responsible for creating. Why has tenure become such a sore subject for administrators and politicians, neither of whom may be scholars or understand what tenure is designed to accomplish. It is always claimed that tenure protects the individual faculty member, which is so, but the real value of the tenure system is the protection of society! By providing means for scholars to pursue their life work under secure conditions, society obtains their services at costs much less than would be required in other arrangements. Without tenure protection we shall see three developments: decline of innovation and disappearance of long-term studies of important unsolved mysteries while faculty "play it safe", unionization of faculty (Call the Teamsters!) with negotiated contracts for the members, and increased costs as new faculty recruits will come only if greatly increased income buys their services. Already faculty hiring practices tend not to be for entry level tenure-track appointments but for appointments of senior funded investigators at greater expense.

5. Knowing the Literature

In today's mad rush to publish something, even as a LPU, it has become possible to publish papers that are not scholarly contributions. The commercial press journals are hungry for "molecular" articles such that editorial scrutiny is minimal in ever so many cases. Among the deficiencies resulting is a lapse in citation of pertinent prior literature.

There is always argument that modern "molecular" research has no precedents, thus that there are no articles for citation. Nonetheless, there is usually pertinent background material that could be cited did the authors care. The genuine expert today reading many articles on their topic of expertise is disappointed that the article is published, but referee and editor review is such that citation of pertinent literature is not had. The inadequacy of editorial process by peers equally uninformed may account for the lapse; perhaps it is too great a burden for authors, editors, and publishers to ensure professional citation of relevant prior literature.

Genuine ignorance may be involved; recall the aphorism "Never attribute to design what may be attributed to incompetence!". On more than one occasion I, as referee, have had to disappoint would-be discoverers of "new" steroids by calling attention to prior works. Admittedly, some of these items appear in out-of-the-way journals, one about isomeric cholesterol chlorohydrins appearing in a

Swedish paper technology journal. The genuine expert would know this.

On this matter I plead guilty to genuine ignorance in two instances. Our 1968 report of the presence of fecal 5ß-sterols in natural waters and sewage and our 1972 report describing use of a reagent for detection of sterol peroxides failed to cite previously published related works, both cases being without my proper awareness at the time of prior relevant literature.

In turn, two examples of our work on steroids have escaped citation, one being from unawareness, one perhaps from other considerations. Our 1989 paper correcting a previously assigned structure of a rearrangement isomer of triamcinolone was followed in 1994 by another structure correction but without citation of our 1989 report.[39] In this case, pure oversight was involved, as a literature search had been made by the authors some years before 1989, but the search had not been continued to 1989. Secondly, our and others' extensive investigations of 1966-1981 of human tissue oxidized cholesterol derivatives (oxysterols)[40,41] have been overlooked, with like material reported as if just discovered.[42]

Disregard of relevant published literature seems to be current fashion for some, fashion based on unrealized ignorance but also on deliberate disregard, thereby to gain unearned credit and advantage. If you seek advantage from your rediscovery of something, why tell the world that the matter was discovered long before by others. It may be human nature to ignore prior matters that are inconvenient. Columbus is said to have visited Iceland in 1485 and learned of Greenland cross the Denmark Strait; should he then have acknowledged the prior Norse discovery of America?

The matter has become so troublesome that a tag "Disregard Syndrome" has been attached to the phenomenon, and suggestions for remedy are being advanced. Among such is demand for a signed statement from authors that an adequate search of the literature has been made![43] Whatever happened to scholarship.

Moreover, deliberate disregard of sound literature unsupportive of a particular partisan concept or political point of view is a darker matter but one that must be recognized where biomedical data may be adduced. A prize example of this wicked bias is the selection of material to be cited supporting controversial social issues of public health, where the premier medical journals *JAMA* and *The New England Journal of Medicine* have allowed selected authors freedom to publish inferior statistics data with assertion that the availability of firearms in the United States is the cause (!) of criminal violence with firearms. The thesis is contentious, but by ignoring all and better data not in support of the thesis one sees how political advocacy has entered into what once was serious biomedical literature.[44]

There is no sound argument for an author to be ignorant of published relevant prior work. Do such lapses reflect ignorance, thus incompetence, or is there a new sort of dishonest PC where prior reports that might cast a shadow on your latest thing are not tolerated. Only the press for instant fame by having no precedents, laziness or incompetence in searching the literature, or wicked design can explain matters. Of course, the genuine scholar knows the literature. The Renascence Man knows even more but nonetheless is disappearing from the scene, along with the Cretaceous dinosaurs.

We are drowning in information but gaining little wisdom. Rediscovery of what is already known is a worthwhile venture and may be had by careful use of the classic tools *Chemical Abstracts* and *Biological Abstracts*, commercial alerting services, and page-by-page search of related papers for background. More recent aids use computer databases and the Science Citation Index of Eugene Garfield's Institute for Scientific Information touted as an aid to ferret out fraud and redundant publications by miscreants. However, the older literature, even landmark papers, does not appear in indices or computer databases and cannot be found by such means. In Molecular Biology the word is that anything older than 5 years is out of date anyway, so throw out your older reprints.

The matter of authorship of science papers long a regular problem now receives new attention. Authorship has always been an issue with those left off, and the order of authors' names, ghost authors, and courtesy authors pose related difficulties. We recognize that all authors are responsible for the whole paper and should be able to discuss or explain the paper as a whole. Present problems are such that the American Medical Association created a remedy in which all authors sign statements vouching for their authorship, what part of the work they did, what relevant financial interests they have.[45]

My own experience with courtesy authors of chemistry papers from my industrial period resulted in three cases of courtesy authorship for persons who did nothing save be there, two requested, the third demanded. The latter case was helpful to my decision to leave the pharmaceutical industry.

We are accustomed to transferring copyright to journals publishing our work. Generally the senior (or submitting) author signs for all, but this has caused problems. In one instance one publisher requires copyright releases from all authors, "including deceased authors" (Churchill Livingstone)!

6. Research Funding

As emphasized here, receipt of federal funds for biomedical research has become absolutely necessary for much serious work. The lure of instant success and world-wide fame available with big federal research grants now intrudes on the simple life of the devoted scientist who must deal with these matters or fail. The ever-present worry of big federal funding has now so permeated science, at least in medical schools, that those without a grant have become nobodies, unless they be related to the Dean or Governor.

Great effort is now expended in arranging funding,

perhaps more so than in actually conducting proposed experimental work, much of which has already been done anyway. The first decision is to determine to which NIH institute to submit a proposal, obviously one with recent improved budgets or rampant media demands for solutions to disorders such as AIDS. Perusal of the composition of the panels reviewing proposals is next important, and one vets lists of Study Section members for names of friend or foe. The right proposal title and selected topic keywords may direct your application to where you want it to go. Pick the right Institute, Study Section, and appealing proposal and funding may be yours. Ignore these steps and by current triage arrangements the proposal is not even read.

Clever scientists recognize these arrangements and formulate their proposals accordingly. It is now necessary to present preliminary data to establish that the proposed work can succeed. It is thus the fashion to submit appealing proposals about work already done, thereby to appear to be more accomplished investigators. One proposes to do work that can be funded, then uses the money for the more creative work. If the innovative work succeeds, renewal applications are also successful; if the new ideas do not work, try again.

Institutions of dependent science now energetically pursue the federal dollar, as must also the individual scientist if he is to have the advantages funded investigators enjoy. We have a shift from state and local support of research as an element in the education of science students to federal funding for goal-oriented applied research. There seems to be no means of return.

Such dependence leads to routine investigations, chink-filling, with imaginative and innovative matters left out from proposals for fear that blue-sky work cannot succeed, that if the ideas expressed appeal to Study Section examiners, the ideas will be exploited by the examiners before the applicant with ideas ever hears of the disposition of his grant application. So, we see standard or routine matters proposed with a flair, with crafty grantsmanship, and ultimately with political influence as congressmen or cabinet members become convinced that their proteges are worthy. Intrusions of ideology, dogma, and PC are now seen in grant proposals.

Several features of the current grants review process are of concern:[46] (1) Competitors of the applicant reviewing the application may not be tolerant or supportive because of the competition involved. (2) There is no genuine expert peer to the applicant on some review panels, resulting in inadequate understanding or appreciation of the application. (3) Contrary remarks about applicant and application are unrefutable, unaccountable. (4) Fashionable projects receive more favorable attention than do truly innovative applications. (5) The limited experience of panel members reduces their capacity to grasp innovative concepts. (6) A "level of enthusiasm" euphemism permits subjectivity to reign in assignment of final scores. (7) The necessity for preliminary data means that the proposed research must already have been done. Awarded funds are then used on new work not necessarily related to the original proposal. (8) Investigators spend too much time and effort on arranging funds and are paid by their grants merely to apply for more money. (9) Lengthy delays between application and review occur, such that new information cannot be included. (10) Travel expenses for panel members are not insignificant but could have been used for research were other means of evaluation of applications available. (11) The required outline of research schedules for several years ignores reality, as changes in directions may occur at any time.

Control of Study Section decisions by cliques or factions intimately involved in research reviewed by the Section is of concern, though officially denied. Rotation onto and off given Study Sections is presumably a decision of the Executive Secretary, now renamed the Scientific Review Administrator, but collusion among parties directs funding to the proper laboratories.

The review of each research proposal by an appointed referee may make or break. An enthusiastic reception by a prominent Study Section member ejaculating "I want to see this funded" settles the matter, as Study Section members regard one another as members of a privileged class with impeccable judgement. Close ranks; he is one of us.

Negative remarks, justified or no, doom a proposal, as also is the case where molecular biology approaches are not proposed. However, imperfections in the NIH system are well known; rejection by one NIH Study Section may be followed by approval by another. Also, the element of pure chance is revealed in the review of 150 NSF research proposals by a second set of referees.[47]

The NIH grants process that has served modern biomedical science so well is now in a new phase; changes in review processes are planned.[48] The NIH tries to discover remedies and alternatives to improve matters and measures to reduce the number of proposals that a Study Section must review. A system of triage is in place; those applications deemed "non-competitive" in an initial administrative preview are simply not evaluated further. There are regular periodic changes in the format of application papers; limits are set on the number of pages; budget arrangements are in revision. New assurances that each current political issue shall be met correctly are required. Above all, "preliminary" evidence of proposal success must be included.

In response to needs for generous federal funding many a research medical school have full-time offices devoted solely to technical aspects of grant applications, of discovering new sources of research funds, and of approval of all funding applications from the school.

7. Federal Science Police

With federal moneys come federal controls, some benign, others not so. Several federal bureaucracies have been established to investigate and regulate aspects of misconduct, fraud, criminal embezzlement, and other deviations in the conduct of government funded research. The Department of Health and Human Services has ORI, and we hear periodically of the good work of the NIH and ORI in managing the problems of science misconduct, if not of deviations.

The history of efforts to manage these problems is intriguing, as there is revealed most of the human failures implicated in science research, research funding, and government regulation. Among such efforts have been those of NIH investigators Walter W. Stewart and Ned Feder who turned their attention to the investigation of fraudulent research, ostensibly research funded by NIH. Their efforts exposed many suspect cases, including the Darsee, Baltimore, Ungar, and Benveniste cases with the different degrees of implied misconduct previously discussed here. However, Stewart and Feder also attracted adverse criticism of their efforts, as their disclosures made some administrators uncomfortable.

The duo had devised a computer program to compare published texts for possible plagiarism, with the claim that they could tell how much of a document had been plagiarized. This effort offended Stephen Oates, University of Massachusetts, Amherst,MA, who complained of their investigation of plagiarism in his biography of Abraham Lincoln.[49] This matter and the DeLuca affair previously described led to the mistreatment of Stewart and Feder as whistleblowers.

The problem is continuing, as the duo reported alleged misconduct of Robert C. Gallo to Kenneth J. Ryan of CRI using NIH letterhead stationary, for which they were reprimanded officially February 1995.[50] Gallo's misconduct case involved his development at NCI of a blood test for the AIDS HIV virus which he patented in 1985 despite an earlier filing by the Pasteur Institute of Paris. The Institute brought suit over the matter, but the suit was settled in 1987 on the assurance by Gallo that his discovery of the HIV virus was independent from that of Luc Montagnier of the Pasteur Institute. Nonetheless, Gallo had received Montagnier's HIV strain and had used it in development of his blood test, later denying these inconvenient facts. Rep. John D. Dingell investigated, and the ORI concluded in 1992 that Gallo "committed science misconduct" by withholding information from the Pasteur Institute, NIH, Department of Justice, and Department of Health and Human Services. However, so overwhelming was the federal government interest in AIDS and the HIV virus that the ORI charges against Gallo were dropped in November 1993, thus according Gallo full protection for his obvious misdeeds.[51] Gallo remains head of his NCI laboratory.

Administrators are reluctant to pursue accusations of misconduct against their prominent faculty and staff members. Moreover, accusations against powerful people whatever their activities may lead to serious difficulties for the accuser. The powerful generally prevail and are cleared of charges as political pressures in their favor develop, witness the Baltimore and DeLuca cases and the Gallo affair just described. The junior accuser is generally the loser, of status, job, and reputation.[52]

Other accounts of such difficulties abound. Faith Fenderson reported to laboratory head Irwin Rose, Fox Chase Cancer Center, Philadelphia, her suspicion that technician Gang Yuan was fabricating data to please Rose, who had outlined what he expected would be results of planned experiments. Shades of William S. Gilbert: "When your Majesty says 'Let a thing be done' it's as good as done - practically, it *is* done". As nothing was done to alter matters Fenderson later reported the suspicion to others outside the laboratory, to a Fox Chase official and to the ORI. Rose fired Fenderson the day he found out her actions; computer files were altered the same day. The ORI concluded that Yuan had fabricated or falsified data; Yuan maintained innocence; Fenderson had difficulty finding further work.[53]

Psychologist Robert L. Sprague in uncovering Steven E. Breuning's misconduct suffered NIH investigation himself and loss of funds. The NIH Council was unable to reconcile comments about the matter. Bruce W. Hollis initially blamed along with Philip W. Lambert over fraudulent work with

Vitamin D was eventually cleared of misconduct but recommends that junior scientists discovering misconduct not blow the whistle, thereby to escape personal suffering. In another incident in which physician Jerome Jacobstein attempted to have the Cornell University Medical College dean investigate alleged misconduct of cardiologist Jeffrey Borer, Jacobstein found that Cornell University was not interested in investigating its medical school.[54]

These problems are neither new nor confined to biomedical science. At the Los Alamos National Laboratory physicist Robert Henson who in 1995 discovered the infamous irregularities in which the Peoples Republic of China gained engineering knowledge of our nuclear W-88 multiple warhead found it extremely difficult to pass his information to laboratory administrators. His message was too damaging to hear. Henson was subsequently fired two times in a laboratory-wide policy to reduce the labor force.[55]

However, there is now a federal government obsession to protect the accuser, the whistleblower, from poor treatment for their disclosures. Even those making false accusations that destroy reputations are protected. Currently about 70% of the misconduct charges investigated by ORI are found without merit, and those charged are exonerated. However, there is a movement for public disclosure of all charges of misconduct whether proven or no, including anonymous accusations. The time-honored opportunity of the accused to confront their accusers is not part of present thinking.

The Office of Special Council (OSC) was created to investigate and prosecute cases of reprisal before the Merit Systems Protection Board (MSPB). Other practices have been devised to protect the whistleblower by shifting matters from science to law courts. We see a Bill of Rights, a Whistleblower Protection Regulation of the ORI, and use of the False Claims Act of 1863 (!) to sue for grievances.[56] It has not been debated whether personal money awards in lawsuits under the False Claims Act creates incentives to whistleblowers or whether their protection but not that of the accused serves modern science.

This amusing history aside, the issue of NIH control of science remains a real one. If your institution gets favored NIH research funds, the ORI becomes your watchdog. By federal regulations, any charge, any complaint, no matter how unfounded or outrageous, signed or anonymous, must be investigated as were it a proven criminal act.

My own single personal experience with this problem was as confidant of two junior faculty members at UTMB charged with misconduct in submitting false or nonexistent data in an NIH grant application. On the day of required investigation following an anonymous, written denunciation charging misconduct, and without prior announcement, the Assistant Dean and other enforcers came to the lab where the alleged miscreants were conducting an experiment. The experiment had to be stopped; a treated animal and data were lost. The criminals had to go directly to their lab notebooks, data collections, and other evidence, all of which was turned over to the Dean, thereby to keep the accused from doctoring the records.

After thorough investigation it was evident that no improper use of data had occurred and that the accused were innocent of the anonymous charges. A letter of exoneration from Assistant Dean Walter J. Meyer III, Scientific Integrity Office, of January 22,1996, marked CONFIDENTIAL mailed to the accused's home address says: "The Committee found no significant evidence to support this claim". This seems to be the best an inquisitional office can do; no apology was forthcoming for high handedness, for ruining an experiment or the reputation of a scientist. The letter of exoneration was posted by the accused on the laboratory bulletin board along with a beautiful picture I provided of SA storm troops burning books in Germany 1933.

The loss of the experiment, inability to access their own data, and a few other indignities caused resentment among the accused. I had been asked for advice by one of them, and I arranged a discussion with the Dean, who explained his fix.

He is required to investigate, no matter the anonymous nature of the complaint. Furthermore, in keeping with efforts to reduce waste, abuse, and fraud UTMB has created a Fraud and Abuse Hotline telephone number 1-(800)-898-7679. Calls are confidential and anonymous.

Additionally, police methods are instituted in the guise of safety. A chemical and biological safety office is needed. Such office can be very helpful to investigators with inadequate knowledge of dangerous biological agents and chemicals, but it is also likely that a dictatorial bureau may result. Just such a case developed at UTMB at a time I was a member of the Chemical Safety Committee. The safety officer decreed by fiat that new regulations requiring eight-page reports on every chemical in the laboratory were to be imposed. What do you have? How much do you have? How are they stored? What provisions have you for spills? How do you dispose of chemicals? Eight pages!

When one faculty member told them he used occasionally one-inch wide asbestos tape to wrap a heated zone on his mass spectrometer a little girl official almost dropped dead. Asbestos! How could you! How much do you use? An inch every year or so. How do you clean up spills? I don't spill.

The list of proscribed chemicals included stearic acid (of soap) and Vitamin D (of white bread), right up there with botulin toxin and nerve gas. My opposition to such ludicrous idiocy, together with more restrained committee members' remarks, altered the proposal. I was removed from the Committee.

These messages of science police are clear; autocratic regulation of science is upon us.

CONCLUSION

Now that I have careered through this mass of material I trust I have not offended nor bored anyone. At least, not too much so. So much more could have been written, but the examples cited appear to characterize the problems perceived. The levity in a few spots is free.

Our science is alive and well, fully capable of discovery of new fundamental information and of advancing at rapid pace to prescribed goals. Our science is likewise capable of handling the many problems here described, if left to science and not to government for resolution. It is merely a matter of being aware of the many aspects of error and misconduct possibly encountered in modern science, and also of the insidious nature of some national movements such as PC to distort things.

Many of the questions arising in our class mentioned here without resolution, definitive description, or other final comment are left in that incomplete state by design, for the interested student to ponder, for the future instructor to treat in his own class.

We must meet these difficulties with stout resolution!

REFERENCES

CHAPTER 1. INTRODUCTION

1. (a) The Diversity Myth, Multiculturism and the Politics of Intolerance at Stanford, Independent Institute, Oakland,CA. (b) S. H. Balch and G. Ricketts, *Wall Street Journal*, June 3,1999, p.A26.
2. *LCGC North America* (*Liquid Chromatography Gas Chromatography*), **20**, 128 (2002).
3. (a) K. W. Woodbury, *The Scientist*, April 26,1999, p.1. (b) T. V. Rajan, *ibid.*, April 29,1996, p.10.
4. Daniel S. Greenberg, The Grant Swinger Papers, Science & Government Report, Washington City,DC, 1982.
5. (a) J. Grisham, *Chemical & Engineering News*, June 21,1999, p.27. (b) *ASBMB NEWS*, **8**, No.5, 5 (1999).
6. D. F. Bloom, J. D. Karp, and N. Cohen, The Ph.D. Process: A Student's Guide to Graduate School in the Sciences, Oxford University Press, New York, 1998.
7. (a) P. Zurer, *Chemical & Engineering News*, June 27,1994, p.35. (b) D. Illman, *ibid.*, October 24,1994, p.38.
8. A. Kornberg, *Biochemistry*, **26**, 6888 (1987).
9. I. Hargittai and J. Maddox, *The Chemical Intelligencer*, **5**, No.2, 53 (1999); I. Hargittai and K. Mullis, *ibid.*, **5**, No.3, 11 (1999).
10. (a) D. Fischer and H. Duerbeck, Hubble: A New Window to the Universe, Copernicus, 1996 and Hubble: Revisited: New Images from the Discovery Machine, Copernicus, 1998. (b) E. Chaisson, The Hubble Wars, Harvard, 1998. (c) C. C. Petersen and J. C. Brandt, Hubble Vision: Further Adventures with the Hubble Space Telescope, Cambridge University Press, 1998.
11. B. Russell, The Autobiography of Bertrand Russell 1872-1914, Little, Brown and Co., Boston, 1967, p.210.

CHAPTER 2. THE COURSE

1. R. C. deL. Milton, S. C. F. Milton, and S. B. H. Kent, *Science*, **256**, 1445 (1992).
2. (a) A Group Of Papers On Medical Writing, Parke, Davis & Co., Detroit, 1957. (b) L. F. Fieser and M. Fieser, Style Guide For Chemists, Reinhold Publishing Corp., New York, 1960. (c) Handbook For Authors Of Papers In The Research Journals Of The American Chemical Society, Washington City, 1965. (d) F. P. Woodford, Scientific Writing For Graduate Students. A Manual On

The Teaching Of Scientific Writing, Rockefeller University Press, New York, 1968. (e) E. J. Huth, How To Write And Publish Papers In The Medical Sciences, ISI Press, Philadelphia, 1982. (f) A. Eisenberg, Effective Technical Communication, McGraw Hill Book Co., Hightstown,NJ, 1982. (g) E. T. Cremmins, The Art Of Abstracting, ISI Press, Philadelphia, 1982. (h) H. B. Michaelson, How To Write And Publish Engineering Papers And Reports, ISI Press, Philadelphia, 1982. (i) R. A. Day, How To Write And Publish A Scientific Paper, ISI Press, Philadelphia, 2nd Ed., 1983. (j) H. E. Ebel, C. Bliefert, and W. E. Russey, The Art Of Scientific Writing: From Student Reports To Professional Publications In Chemistry And Related Fields, VCH Publishers, New York, 1987.

3. (a) T. Mill, D. G. Hendry, and H. Richardson, *Science*, **207**, 886 (1980). (b) R. M. Baxter and J. H. Carey, *Nature*, **306**, 575 (1983). (c) W. J. Cooper and R. G. Zika, *Science*, **220**, 711 (1983). (d) R. G. Petasne and R. G. Zika, *Nature*, **325**, 516 (1987).

4. (a) S. F. Arnold, D. M. Klotz, B. M. Collins, P. M. Vonier, L. J. Guilette, and J. A. McLachlan, *Science*, **272**, 1489 (1996). (b) J. A. McLachlan, *Science*, **277**, 462 (1997). (c) *Chemical & Engineering News*, June 10,1996, p.6; June 16,1996, p.31; July 28,1997, pp.9-10.

5. (a) L. Carroll, The Hunting of the Snark, MacMillan Co., London, 1886. (b) L. Carroll, The Hunting of the Snark, in M. Gardner, The Annotated Snark, Bramhall House, New York, 1962.

6. A. U. Khan, *Journal of Physical Chemistry*, **80**, 2219 (1976).

7. D. M. Gibson and T. S. Ingebritsen, *Life Sciences*, **23**, 2619 (1978).

8. M. Zasloff and G. Felsenfeld, *Biochemistry*, **16**, 5135 (1977).

9. J. W. Peters, J. N. Pitts, I. Rosenthal, and H. Fuhr, *Journal of the American Chemical Society*, **94**, 4348 (1972).

10. (a) E. K. Hodgson and I. Fridovich, *Biochemistry*, **13**, 3811 (1974). (b) J. W. Peters, P. J. Bekowies, A. M. Winer, and J. N. Pitts, *Journal of the American Chemical Society*, **97**, 3299 (1975).

11. I. Rosenthal, *Israel Journal of Chemistry*, **13**, 86 (1975).

12. (a) A. M. Held and J. K. Hurst, *Biochemical and Biophysical Research Communications*, **81**, 878 (1978). (b) Y. Ushijima and M. Nakano, *ibid.*, **93**, 1232 (1980).

13. (a) P. B. Merkel and D. R. Kearns, *Journal of the American Chemical Society*, **94**, 7244 (1972). (b) J. R. Hurst, J. D. McDonald, and G. B. Schuster, *ibid.*, **104**, 2065 (1982). (c) M. A. J. Rodgers and P. T. Snowden, *ibid.*, **104**, 5541 (1982). (d) M. A. J. Rodgers, *Photochemistry & Photobiology*, **37**, 99 (1983).

14. H. L. Pahl and P. A. Baeuerie, *Bioessays*, **16**, 497 (1994).

15. L. B. Clerch and D. J. Massaro, Editors, Oxygen, Gene Expression, and Cellular Function, Marcel Dekker Inc., New York City, 1997.

16. S. Legrand-Poels, M. Hoebeke, D. Vaira, B. Rentier, and J. Piette, *Journal of Photochemistry and Photobiology B: Biol*, **17**, 229

(1993).

17. (a) S. I. Liochev and I. Fridovich, *Archives of Biochemistry*, **337**, 115 (1997); *Free Radical Biology and Medicine*, **25**, 926 (1998). (b) I. Fridovich, *Journal of Biological Chemistry*, **272**, 18515 (1997). (c) I. Heiser, A. Muhr, and E. F. Elstner, *Zeitschrift für Naturforschung C*, **53**, 9 (1998). (d) Y. Li, H. Zhu, P. Kuppusamy, V. Roubaud, J. L. Zweier, and M. A. Trush, *Journal of Biological Chemistry*, **273**, 2015 (1998). (e) M. P. Skatchkov, D. Sperling, U. Hink, A. Mülsch, D. G. Harrison, I. Sindermann, T. Meinertz, and T. Münzel, *Biochemical and Biophysical Research Communications*, **254**, 319 (1999). (f) I. B. Afanas'ev, E. A. Ostrachovitch, and L. G. Korkina, *Archives of Biochemistry and Biophysics*, **366**, 267 (1999). (g) I. B. Afanas'ev, E. V. Mikhal'chik, E. A. Ostrachovitch, and L. G. Korkina, *Free Radical Biology and Medicine*, **29**, Suppl.1, S21 (2000). (h) M. Janiszewski, H. P. Souza, X. Liu, M. A. Pedro, J. Zweier, and J. Laurindo, *ibid.*, **29**, Suppl.1, S119 (2000); M. Janiszewski, H. P. Souza, X. Liu, M. A. Pedro, J. L. Zweier, and F. T. M. Laurindo, *ibid.*, **32**, 446 (2002).

18. G. Cighetti, S. Denbiasi, P. Ciuffreda, and P. Allevi, *Free Radical Biology and Medicine*, **25**, 818 (1998).

19. (a) *Wall Street Journal*, June 30,1969; (b) *New York Times*, September 12,1969.

20. D. Steinberg, *Journal of Biological Chemistry*, **272**, 20963 (1997).

21. J. H. Hildebrand, *Science*, **168**, 1397 (1970).

22. J. D. Bernal, P. Barnes, I. A. Cherry, and J. L. Finney, *Nature*, **224**, 393 (1989).

23. 44th National Colloid Symposium on Anomalous Water, June 1970; *Journal of Colloid and Interface Science*, **36**, 434 (1971).

24. B. V. Derjaguin and N. V. Churaev, *Kolloidnyi Zhurnal*, **35**, 814 (1973); *Nature*, **244**, 430 (1973).

25. F. Franks, Polywater, MIT Press, Cambridge, 1981.

26. (a) L. Allen, *New Scientist*, **59**, No.859, 376 (1973). (b) M. P. Gingold, *Bulletin de la Société Chimique de France*, **40**, 1629 (1973).

27. *New Scientist*, **122**, No.1670, 31 (1989).

28. E. W. Lang and H.-D. Lüdemann, *Angewandte Chemie International Edition*, **21**, 315 (1982); *Angewandte Chemie*, **94**, 351 (1982).

29. (a) N. Pugliano and R. J. Saykally, *Science*, **257**, 1937 (1992). (b) J. D. Cruzan, L. B. Braly, K. Liu, M. G. Brown, J. G. Loeser, and R. J. Saykally, *ibid.*, **271**, 59 (1996). (c) K. Liu, M. G. Brown, J. D. Cruzan, and R. J. Saykally, *ibid.*, **271**, 62 (1996).

30. (a) A. Haim, *Journal of the American Chemical Society*, **114**, 8384 (1992). (b) F. M. Menger, *Journal of Organic Chemistry*, **56**, 6251, 6960 (1991). (c) F. M. Menger and A. Haim, *Nature*, **359**, 666 (1992). (d) P. Zurer, *Chemical & Engineering News*, October

26,1992, p.6.

31. R. Breslow, *Accounts of Chemical Research*, **24**, 317 (1991).
32. C. Gutierrez, Z.-S. Guo, W. Burhans, M. L. DePanphilis, J. Farrell-Towt, and G. Ju, *Science*, **240**, 1202 (1988).
33. G. P. Studzinski, *Science*, **240**, 1202 (1988).
34. (a) W. Thompson and G. MacDonald, *Journal of Biological Chemistry*, **253**, 2712 (1978); **254**, 3311 (1979). (b) W. Thompson and R. T. Zuk, *ibid.*, **258**, 9623 (1983).
35. (a) C. Milanese, N. E. Richardson, and E. L. Reinherz, *Science*, **231**, 1118 (1986); **234**, 1056 (1986). (b) C. Milanese, R. E. Siliciano, R. E. Schmidt, J. Ritz, N. E. Richardson, and E. L. Reinherz, *Journal of Experimental Medicine*, **163**, 1583 (1986). See also (c) B. J. Culliton, *Science*, **234**, 1069 (1986); E. Collins, *Nature*, **324**, 197 (1986).
36. (a) J. Sakai, A. Nohturfft, J. L. Goldstein, and M. S. Brown, *Journal of Biological Chemistry*, **273**, 5785 (1998). (b) A. Nohturfft, M. S. Brown, and J. L. Goldstein, *ibid.*, **273**, 17243 (1998). (c) J.-t. Pai, O. Guryev, M. S. Brown, and J. L. Goldstein, *ibid.*, **273**, 26138 (1998).
37. (a) R. J. Gullis and C. E. Rowe, *Biochemical Journal*, **148**, 197 (1975). (b) R. J. Gullis and C. E. Rowe, *Biochemistry Society Transactions*, **1**, 849 (1973). (c) R. J. Gullis and C. E. Rowe, *Biochemical Journal*, **148**, 197, 557, 567 (1975). (d) R. J. Gullis and C. E. Rowe, *Journal of Neurochemistry*, **26**, 1217 (1976). (e) R. J. Gullis and C. E. Rowe, *FEBS Letters*, **67**, 256 (1976).
38. (a) R. J. Gullis, J. Traber, and B. Hamprecht, *Nature*, **256**, 57 (1975). (b) R. J. Gullis, J. Traber, K. Fischer, C. Buchen, and B. Hamprecht, *FEBS Letters*, **59**, 74 (1975). (c) R. J. Gullis, C. Buchen, L. Moroder, E. Wünsch, and B. Hamprecht, in H. W. Kosterlitz, Editor, Opiates and Endogenous Opioid Peptides, Elsevier, Amsterdam, 1976, pp.143-151. (d) M. Brandt, R. J. Gullis, K. Fischer, C. Buchen, B. Hamprecht, L. Moroder, and E. Wünsch, *Nature*, **262**, 311 (1976).
39. (a) B. Hamprecht, *Nature*, **265**, 764 (1977); (b) R. J. Gullis, *ibid.*, **265**, 764 (1977).
40. (a) M. H. Rosner, R. J. De Santo, H. Arnheiter, and L. M. Staudt, *Cell*, **69**, 724 (1992). (b) M. H. Rosner, M. A. Vigano, P. W. J. Rigby, H. Arnheiter, and l. M. Staudt, *Science*, **253**, 144 (1991); **257**, 147 (1992).
41. G. Zadel, C. Eisenbraun, G.-J. Wolff, and E. Breitmaier, *Angewandte Chemie International Edition*, **33**, 454 (1994); *Angewandte Chemie*, **106**, 460 (1994).
42. (a) C. A. Mead and A. Moscowitz, *Journal of the American Chemical Society*, **102**, 7301 (1980). (b) L. D. Barron, *Science*, **266**, 1491 (1994).
43. E. Breitmaier, *Angewandte Chemie International Edition*, **33**, 1207

(1994).

44. S. C. Sinha, C. F. Barbas, and R. A. Lerner, *Proceedings of the National Academy of Sciences USA*, **95**, 14603 (1998).

45. (a) M. Spector, S. O'Neal, and E. Racker, *Journal of Biological Chemistry*, **255**, 5504, 8370 (1980). (b) E. Racker and M. Spector, *Science*, **213**, 303 (1981). (c) M. Spector, R. B. Pepinsky, V. M. Vogt, and E. Racker, *Cell*, **25**, 9 (1981). (d) A. Rephaeli, M. Spector, and E. Racker, *Journal of Biological Chemistry*, **256**, 6069 (1981).

46. (a) E. Racker, *Science*, **213**, 1313 (1981). (b) V. M. Vogt, R. B. Pepinsky, and E. Racker, *Cell*, **25**, 827 (1981). (c) E. Racker, *Science*, **222**, 232 (1983). (d) E. Racker, *Federation Proceedings*, **42**, 2899 (1983). (e) E. Racker, *Nature*, **339**, 91 (1989).

47. R. Breslow and M. P. Mehta, *Journal of the American Chemical Society*, **108**, 2485, 6417, 6418 (1986).

48. (a) R. Breslow, *Chemical & Engineering News*, December 8,1986, p.2, p.6; May 11,1987, p.2. (b) R. Breslow, *Journal of the American Chemical Society*, **109**, 1605 (1987).

49. (a) D. R. VanDeripe, G. B. Hoey, W. R. Teeters, and T. R. Tusing, *Journal of the American Chemical Society*, **88**, 5365 (1966). (b) L. R. Axelrod and P. N. Rao, *ibid.*, **89**, 5313 (1967).

50. G. Orwell, "Politics and the English Language", in The Orwell Reader Fiction, Essays, and Reportage by George Orwell, Harcourt, Brace & World Inc., New York, 1956, pp.355-366.

CHAPTER 3. SOME CURRENT PROBLEMS

1. J. Horgan, The End of Science: Facing the Limits of Knowledge in the Twilight of the Scientific Age, Addison Wesley, New York, 1996.

2. (a) M. J. Behe, Darwin's Black Box: The Biochemical Challenge to Evolution, Free Press, New York, 1996. (b) W. A. Dembski, Mere Creation: Science Faith and Intelligent Design, InterVarsity Press. (c) W. A. Dembski, Intelligent Design: The Bridge Between Science and Theology, InterVarsity Press, 1999. (d) J. Wells, Icons of Evolution. Science or Myth? Why Much of What We Teach About Evolution Is Wrong, Regnery Publishing Inc., Washington City,DC. (e) C. G. Hunter, Darwin's God: Evolution and the Problem of Evil, Regnery Publishing Inc., Washington City,DC. (f) M. J. Behe, W. A. Dembski, and S. C. Meyer, Science and Evidence for Design in the Universe, Ignatius Press, 2000. (g) J. Sarfati, Refuting Evolution. A Handbook for Students, Parents, and Teachers Countering the Latest Arguments for Evolution, Vision Forum, San Antonio,TX.

3. R. Selzer, *Chemical & Engineering News*, April 24,1995, pp.52-53.

4. M. Schwartz and Task Force on Bias-Free Language, Guidelines for Bias-Free Writing, Indiana University Press, 1995.

5. A. Ross, Strange Weather, Culture, Science, and Technology in the Age of Limits, Verspo Books, 1991.
6. (a) N. K. Hayles, Chaos Bound: Orderly Disorder in Contemporary Literature and Science, Cornell University Press, Ithaca,NY, 1990; (b) N. K. Hayles, Chaos and Order: Complex Dynamics in Literature and Science, University of Chicago Press, Chicago,IL, 1991.
7. P. R. Gross and N. Levitt, Higher Superstition: The Academic Left and Its Quarrels with Science, Johns Hopkins University Press, Baltimore, 1994.
8. N. Koertge, A House Built on Sand: Exposing Postmodernist Myths About Science, Oxford University Press, Oxford, 1998.
9. C. H. Sommers, *Wall Street Journal*, July 10,1995.
10. In the series Race, Gender, and Science the Indiana University Press, Bloomington,IN has issued: (a) N. Tuana, Editor, Feminism and Science, 1989. (b) N. Tuana, The Less Noble Sex. Scientific Religious and Philosophical Conceptions of Woman's Nature, 1993. (c) B. B. Spanier, Im/Partial Science, Gender Ideology in Molecular Biology, 1995. (d) J. Barr and J. Birke, Common Science? Women, Science, and Knowledge, 1998. (e) S. Harding, Is Science Multicultural? Postcolonialisms, Feminisms, and Epistemologies, 1998.
11. (a) R. J. Herrnstein and C. Murray, The Bell Curve: Intelligence and Class Structure in American Life, Free Press, New York, 1996. (b) J. P. Rushton, Race, Evolution and Behavior, Transaction Publishers, New Brunswick,NJ, 1994.
12. J. P. Rushton, *The Scientist*, October 3,1994, p.13.
13. (a) Daubert v. Merrill Dow Pharmaceuticals Inc., 1993; *Chemical & Engineering News*, December 14,1998, p.31. (b) D. L. Faigman, Legal Alchemy: The Use and Misuse of Science in the Law, W. H. Freeman and Co., San Francisco, 1999.
14. (a) A. González Thompson and R. I. Charles, Addison-Wesley Secondary Math: An Integrated Approach: Focus on Algebra, Addison-Wesley Publishing Co., Menlo Park,CA., 1996. (b) M. Gardner, J. Trentacosta, and M. J. Kenney, Multicultural and Gender Equity in the Mathematics Classroom: The Gift of Diversity, National Council of Teachers of Mathematics, 1997.
15. C. J. Sykes, Dumbing Down Our Kids, St. Martin's Press, New York, 1995.
16. W. Osler, Aequanimitas With Other Addresses to Medical Students, Nurses and Practitioners of Medicine, Blakiston Co., New York, 3rd Ed., 1932, pp.117-129.
17. (a) K. J. Arrow, B. C. Eaves, and I. Olkin, Editors, Education in a Research University, Stanford University Press, Stanford,CA, 1996. (b) J. R. Cole, E. G. Barber, and S. R. Graubard, Editors, The Research University in A Time of Discontent, Johns Hopkins

University Press, Baltimore,MD, 1994.

18. Y. Furukawa, Inventing Polymer Science: Staudinger, Carothers, and the Emergence of Macromolecular Chemistry, University of Pennsylvania Press, Philadelphia, 1998.
19. (a) P. Karlson, *Perspectives in Biology and Medicine*, **6**, 203 (1963). (b) P. Karlson, D. Doenecke, and C. E. Sekeris, in M. Florkin and E. H. Stotz, Editors, Comprehensive Biochemistry, Volume 25. Regulatory Functions - Mechanisms of Hormone Action, Elsevier Scientific Publishing Co., Amsterdam, 1975, pp.1-10.
20. (a) H. Selye, *Proceedings of the Society of Experimental Biology*, **4**, 116 (1941); *Endocrinology*, **30**, 437 (1942). (b) For other evidence, see also *Steroids*, **64**, 3-175 (1999); D. W. Braun, L. B. Hendry, and V. B. Mahesh, *Journal of Steroid Biochemistry and Molecular Biology*, **52**, 113 (1995); M. Wehling, *Journal of Molecular Medicine*, **73**, 439 (1995); and M.-T. Sutter-Dub, *Steroids*, **67**, 77 (2002).
21. W. Osler, Aequanimitas With Other Addresses to Medical Students, Nurses and Practitioners of Medicine, Blakiston Co., New York, 3rd Ed., 1932, p.375.
22. J. Watson and F. Crick, *Nature*, **171**, 737 (1953).
23. N. Wade, The Nobel Duel: Two Scientists' 21-Year Race to Win the World's Most Coveted Research Prize, Anchor Press/Doubleday, Garden City,NY, 1981.
24. R. K. Merton and R. Lewis, *Impact of Science on Society*, **21**, 151 (1971).
25. (a) E. Burgos-Debray, Editor, I, Rigoberta Menchú: An Indian Woman in Guatamala, Verso, 1984. (b) D. Stoll, Rigoberta Menchú and the Story of All Poor Guatamalans, Westview. (c) S. Schwartz, *Wall Street Journal*, December 28,1998, p.A14. (d) John Leo, *U.S. News & World Reports*, January 25,1999, p.17. (e) P. Canby, *New York Review of Books*, April 8,1999, pp.28-33.
26. (a) M. A. Bellesiles, Arming America: The Origins of a National Gun Culture, Alfred A. Knopf, New York, 2000. (b) E. S. Morgan, *New York Review of Books*, October 19,2000, p.30, p.32. (c) C. Cramer, *American Rifleman*, January 2001, p.66. (d) K. A. Stassel, *Wall Street Journal*, April 9,2001; February 22,2002. (e) J. Williamson, *New Gun Week*, May 1,2001, p.13. (f) N. Knox, *Shotgun News*, March 1,2001, p.11. (g) M. Korda, *American Rifleman*, November 2001, p.58. (h) D. Workman, *New Gun Week*, January 1,2002, p.9; February 10,2002, p.12.
27. *The Scientist*, October 2,1995, p.3; November 23,1998, p.6, p.13.
28. E. Davenas, F. Beauvais, J. Amara, M. Oberbaum, B. Robinzon, A. Miadonna, A. Tedeschi, B. Pomeranz, P. Fortner, P. Belon, J. Sainte-Laudy, B. Poitevin, and J. Benveniste, *Nature*, **333**, 816 (1988).
29. M. Sidoli, *Journal of Analytical Psychology*, **41**, No.2, 165 (1996).

30. *U.S. News & World Reports*, December 7,1998, p.18.
31. "Sacred Writings", Harvard Classics, P. F. Collier & Son Co., New York, 1910, Vol.44, p.43.
32. "The Apology, Phædo and Crito of Plato", Harvard Classics, P. F. Collier & Son Co., New York, 1909, Vol.2, p.112.
33. J. Lienhard, Engines of Our Ingenuity No.583, University of Houston, Houston,TX.
34. P. Boehner, Philosophical Writings, A Selection, William of Ockham, Bobbs-Merrill Co., Indianapolis/New York, 1964, pp.xx-xxi.
35. A. Koyré and I. B. Cohen, Editors, Isaac Newton's Philosophiæ Naturalis Principia Mathematica, Harvard University Press, Cambridge, 1972, Vol.II, p.550.
36. B. Russell, The Autobiography of Bertrand Russell 1944-1969, Simon and Schuster, New York, 1969, pp.71-72.
37. Complete Poems of Robert Frost 1949, Henry Holt and Co., New York, 1949, p.495.
38. A. M. Grimwade, *The Scientist*, February 1,1999, p.12.
39. S. Wilkinson, *Chemical & Engineering News*, December 21,1998, pp.41-42.
40. K. S. Thompson, *American Scientist*, **82** 508 (1994).

CHAPTER 4. BEWARE THE MEANING OF WORDS

1. D. P. Hajjar and M. E. Haberland, *Journal of Biological Chemistry*, **272**, 22975 (1997).
2. S. Borman, *Chemical & Engineering News*, June 19,1995, pp.23-32.
3. S. Levin, ΦBK *Key Reporter*, **52** [3], 3 (1987).
4. R. Langreth, *Wall Street Journal*, May 7,1999, p.B1.
5. D. F. Cully, H. S. Ip, and G. A. M. Cross, *Cell*, **42**, 173 (1985).
6. A. Kotyk, Quantities, Symbols, Units, and Abbreviations in the Life Sciences, Humana Press, Totowa,NJ, 1999.
7. A.B. Smith, N. Kanoh, H. Ishiyama, and R. A. Hartz, *Journal of the American Chemical Society*, **122**, 11254 (2000); S. Borman, *Chemical & Engineering News*, November 13,2000, p.12.
8. A. T. Wilson and M. Calvin, *Journal of the American Chemical Society*, **77**, 5948 (1955).
9. D. A. McGurk, J. Frost, E. J. Eisenbraun, K. Vick, W. A. Drew, and J. Young, *Journal of Insect Physiology*, **12**, 1435 (1966).
10. D. M. Smith, *Chemistry & Industry*, 519 (1954).
11. (a) H. Walpole, The Three Princes of Serendip, 1754. (b) T. J. Sommer, *The Scientist*, February 1,1999, p.13.
12. (a) Yu. Ts. Oganessian, "Synthesis and Radioactive Properties of the Heaviest Nuclei", The Robert A. Welch Foundation 41st Conference on Chemical Research. The Transactinide Elements, Houston,TX,

1997, pp.347-367. (b) Yu. T. Oganessian, A. V. Yeremin, A. G. Popeko, S. L. Bogomolov, G. V. Buklanov, M. L. Chelnokov, V. I. Chepigin, B. N. Gikal, V. A. Gorshkov, G. G. Gulbekian, M. G. Itkis, A. P. Kabachenko, A. Yu. Lavrentev, O. N. Malyshev, J. Rohac, R. N. Sagaidak, S. Hofmann, S. Saro, G. Giardina, and K. Morita, *Nature*, **400**, 242 (1999). (c) R. Stone, *Science*, **283**, 474 (1999).

13. (a) C. Thomas, *Austin American-Statesman*, Austin,TX, February 27,1999, p.A11. (b) D. Patai, Professing Feminism: Cautionary Tales from the Strange World of Women's Studies, Basic Books, New York, 1994. (c) C. A. MacKinnon, Feminism Unmodified: Discourses on Life and Law, Harvard University Press, Cambridge, 1987.

14. *U.S. News & World Reports*, August 10,1998, p.12.

15. J. G. Bednorz and K. A. Müller, *Zeitschrift für Physik B*, **64**, 189 (1986).

16. M. K. Wu, J. R. Ashburn, C. J. Torng, P. H. Hor, R. L. Meng, L. Gao, Z. J. Huang, Y. Q. Wang, and C. W. Chu, *Physical Review Letters*, **58**, 908 (1987).

17. W. Pigman, *Chemical & Engineering News*, February 16,1953, p.652.

18. W. E. Moerner, *Accounts of Chemical Research*, **29**, 563 (1996).

19. *Chemical & Engineering News*, September 29,1975, p.27; January 27,1997, p.72; February 17,1997, p.72; March 10,1997, p.200.

20. J. W. Young, Totalitarian Language. Orwell's Newspeak and Its Nazi and Communist Origins, University of Virginia Press, Charlottesville,VA, 1991.

21. "The Apology, Phædo and Crito of Plato", Harvard Classics, P. F. Collier & Son Co., New York, 1909, Vol.2, p.5.

22. (a) Z. Medvedev, The Rise and Fall of T. D. Lysenko, Columbia University Press, 1969. (b) D. Joravsky, The Lysenko Affair, Harvard University Press/University Chicago Press, 1970/1986. (c) V. N. Soyfer, Lysenko and the Tragedy of Soviet Science, Rutgers University Press, New Brunswick,NJ, 1994.

23. L. N. Trut, *American Scientist*, **87**, 160 (1999).

24. (a) *Revue Roumaine de Chimie*, **34**, No.1-No.12 (1989). (b) W. Lepkowski, *Chemical & Engineering News*, February 12,1990, pp.21-22. (c) E. Behr, Kiss the Hand You Cannot Bite. The Rise and Fall of the Ceauşescus, Villard Books, New York, 1991.

25. (a) G. Orwell, 1984, Harcourt, Brace and Company, Inc., 1949, pp.227-237. (b) G. Orwell, "Appendix. The Principles of Newspeak", in The Orwell Reader Fiction, Essays, and Reportage by George Orwell, Harcourt, Brace & World Inc., New York, 1956, pp.409-419.

26. R. Dooling, *Wall Street Journal*, June 29,1998.

27. L. Carroll, Through the Looking Glass and What Alice Found There, Avenel Books, New York, p.124.

28. (a) R. Dooling, *Wall Street Journal*, January 29,1999. (b) J. Leo, *U.S. News & World Reports*, February 8,1999, p.15.

29. L. Bader, *Wall Street Journal*, February 3,1999, p.A23.

30. L. L. Smith, *Lipids*, **31**, 453 (1996).

31. (a) T. Monmaney, *Austin American-Statesman*, Austin,TX, January 16,1999. (b) M. Fumento, *Wall Street Journal*, January 21,1999, p.A18. (c) J. Couzin, *U.S. News & World Reports*, January 25,1999, p.61.

32. P. B. Horton, *Intellect*, **105**, 159 (1976).

33. L. L. Smith, *Journal of Irreproducible Results*, in press!

34. (a) R. Dagani, *Chemical & Engineering News*, June 14,1999, p.6. (b) M. Jacoby, *ibid.*, August 6,2001, p.10.

CHAPTER 5. MODERN REVOLUTIONARY AND MIRACLE SCIENCE

1. (a) W. Broad and N. Wade, Betrayers of Truth: Fraud and Deceit in the Halls of Science, Simon and Schuster/Touchstone, New York, 1982. (b) A. Kohn, False Prophets. Fraud and Error in Science and Medicine, Basil Blackwell Inc., Oxford/New York, 1986. (c) M. C. LaFollette, Stealing Into Print: Fraud, Plagiarism, and Misconduct in Scientific Publishing, University of California Press, Berkeley,CA, 1992. (d) R. Bell, Impure Science: Fraud, Compromise and Political Influence in Scientific Research, John Wiley & Sons, New York, 1992. (e) D. J. Miller and M. Hersen, Research Fraud in the Behavioral and Biomedical Sciences, John Wiley & Sons, New York, 1992. (f) Responsible Science: Ensuring the Integrity of the Research Process, National Academy of Science Press, Vol.I, 1992. (g) C. Crossin, Tainted Truth: The Manipulation of Fact in America, Simon and Schuster, New York, 1994.

2. A. Koestler, The Case of the Midwife Toad, Pan Books, London, 1971.

3. J. Sapp, Where the Truth Lies: Franz Moewus and the Origins of Molecular Biology, Cambridge University Press, Cambridge, 1990.

4. *Chemical & Engineering News*, May 6,1991, pp.43-44.

5. (a) A. T. Petit and P. L. Dulong, *Annales de Chimie et de Physics*, **10**, 395 (1819). (b) *Chemical & Engineering News*, June 15,1987, p.3.

6. (a) J. H. Wotiz and S. Rudofsky, *Journal of Chemical Education*, **59**, 23 (1982). (b) J. H. Wotiz, Editor, The Kekulé Riddle: A Challenge for Chemists and Psychologists, Cache River Press, Vienna,IL, 1993.

7. (a) A. J. Rocke, The Quiet Revolution: Herman Kolbe and the

Science of Organic Chemistry, University of California Press, Berkeley, 1993. (b) *Chemical & Engineering News*, August 23,1993, pp.20-21.

8. (a) G. L. Geison, The Private Science of Louis Pasteur, Princeton University Press, Princeton,NJ, 1995. (b) P. Debré (E. Forster, translator), Louis Pasteur, Johns Hopkins University Press, Baltimore,MD, 1998. (c) *Chemical & Engineering News*, August 7,1995, pp.32-33.

9. R. A. Fisher, *Annales of Science*, **1**, 115 (1936).

10. *Proceedings of the National Academy of Sciences USA*, **86**, 9063 (1989).

11. (a) R. Blondlot, N Rays, Longman Green and Co., London, 1905. (b) I. Frith, *New Scientist*, **44**, No.681, 642 (1969).

12. (a) W. Broad and N. Wade, Betrayers of Truth: Fraud and Deceit in the Halls of Science, Simon and Schuster/Touchstone, New York, 1982, p.23, pp.33-36, pp.227-228. (b) D. Goodstein, *The Scientist*, March 2,1992, p.11, p.13; *American Scientist*, **89**, No.1, 54 (2001).

13. (a) D. Freeman, Margaret Mead and Samoa. The Making and Unmaking of An Anthropological Myth, Harvard University Press, Cambridge,MA, 1983. (b) D. Freeman, The Fateful Hoaxing of Margaret Mead. A Historical Analysis of Her Samoan Research, Westview Press, Boulder,CO, 1998. (c) D. A. Price, *Wall Street Journal*, March 3,1999, p.A17. (d) H. Hellman, Great Feuds in Science: Ten of the Liveliest Disputes Ever, John Wiley and Sons, New York, 1998.

14. F. Wood, Did Marco Polo Go To China?, Westview Press, Boulder,CO, 1998.

15. (a) M. Lasswell, "The 9,400-Year-Old Man", *Wall Street Journal*, January 8,1999, p.W11. (b) D. H. Thomas, Skull Wars: Kennewick Man, Archeology, and the Battle for Native American Identity, Basic Books, New York, 2000. (c) J. C. Chatters, Ancient Encounters: Kennewick Man and the First Americans, Simon & Schuster, New York, 2001.

16. C. W. Petit, *U.S. News & World Reports*, October 12,1998, pp.56-64.

17. L. J. Schiffer, *Wall Street Journal*, January 29,1999, p.A15.

18. E. M. McMillan and P. H. Abelson, *Physics Review*, **57**, 1185 (1940).

19. G. Ungar, D. M. Desiderio, and W. Parr, *Nature*, **238**, 198 (1972).

20. W. W. Stewart, *Nature*, **238**, 202 (1972).

21. G. Ungar, *Nature*, **238**, 209 (1972).

22. E. Davenas, F. Beauvais, J. Amara, M. Oberbaum, B. Robinzon, A. Miadonna, A. Tedeschi, B. Pomeranz, P. Fortner, P. Belon, J. Sainte-Laudy, B. Poitevin, and J. Benveniste, *Nature*, **333**, 816 (1988).

23. J. Maddox, J. Randi, and W. W. Stewart. *Nature*, **334**, 287 (1988).

24. J. Benveniste, *Nature*, **334**, 291 (1988).
25. S. J. Hirst, N. A. Hayes, J. Burridge, F. L. Pearce, and J. C. Foreman, *Nature*, **366**, 525 (1993).
26. J. Benveniste, B. Ducot, and A. Spira and F. A. C. Wiegant, *Nature*, **370**, 322 (1994).
27. (a) J. Baggott, Perfect Symmetry: The Accidental Discovery of a New Form of Carbon, Oxford University Press, 1994. (b) H. Aldersey-Williams, The Most Beautiful Molecule: The Discovery of the Buckyball, John Wiley & Sons, New York, 1995. (c) R. Baum, *Chemical & Engineering News*, January 6,1997, pp.29-33.
28. (a) S. H. Friedman, D. L. Decamp, R. P. Sijbesma, G. Srdanov, F. Wudl, and G. L. Kenyon, *Journal of the American Chemical Society*, **115**, 6506 (1993). (b) R. Sijbesma, G. Srdanov, F. Wudl, J. A. Castoro, C. Wilkins, S. H. Friedman, D. L. Decamp, and G. L. Kenyon, *ibid.*, **115**, 6510 (1993).
29. H. Tokuyama, S. Yamago, E. Nakamura, T. Shiraki, and Y. Sugiura, *Journal of the American Chemical Society*, **115**, 7918 (1993).
30. A. Sygula and P. W. Rabideau, *Journal of the American Chemical Society*, **120**, 12666 (1998).
31. R. C. deL. Milton, S. C. F. Milton, and S. B. H. Kent, *Science*, **256**, 1445 (1992).
32. (a) K. Gademan, M. Ernst, D. Hoyer, and D. Seebach, *Angewandte Chemie International Edition*, **38**, 1223 (1999). (b) *Chemical & Engineering News*, June 16,1997, p.32; May 4,1998, p.56; June 28,1999, p.27; July 26,1999, p.41.
33. R. A. Skelton, T. E. Marston, and G. D. Painter, The Vinland Map and the Tartar Relation, Yale University Press, New Haven,CT, 1965.
34. (a) W. C. McCrone, *Accounts of Chemical Research*, **23**, 77 (1990). (b) W. C. McCrone, Judgment Day for the Shroud of Turin, Prometheus Books, Amherst,NY, 1999. (c) H. E. Gove, From Hiroshima to the Iceman: The Development and Applications of Accelerator Mass Spectrometry, Institute of Physics Publishing, Bristol, 1999. (d) V. M. Bryant, *Biblical Archeology Review*, **26**, No.8, 36 (2000).
35. T. A. Cahill, R. N. Schwab, B. H. Kusko, R. A. Eldred, G. Möller, D. Dutschke, and D. L. Wick, *Analytical Chemistry*, **59**, 829 (1987).
36. (a) H. Kersten and E. R. Gruber, The Jesus Conspiracy: The Turin Shroud & The Truth About the Resurrection, Barnes & Noble, New York, 1995. (b) R. Hoare, The Turin Shroud Is Genuine, Souvenir Publishers, 1999. (c) L. Picknett and C. Prince, Turin Shroud: In Whose Image? The Truth behind the Centuries-Long Conspiracy of Silence, Harper Collins, 1995. (d) I. Wilson, The Blood and the Shroud, Free Press, New York, 1998. (e) J. Nickell, Inquest on the Shroud of Turin: Latest Scientific Findings, Prometheus Books,

Amherst,NY, 1998. (f) M. Whanger and A. Whanger, The Shroud of Turin: An Adventure of Discovery, Providence House, Franklin,TN, 1998. (g) A. Danin, U. Baruch, A. Whanger, and M. Whanger, Flora of the Shroud of Turin, Missouri Botanical Garden Press, St Louis,MO, 1999.

37. A. Eschenmoser, "Toward A Chemical Etiology of the Natural Nucleic Acids' Structure", Proceedings of the Robert A. Welch Foundation 37th Conference on Chemical Research. 40 Years of the Double Helix, Houston,TX, 1994, pp.201-235.

38. D. Nelkin, Selling Science: How the Press Covers Science and Technology, W. H. Freeman and Co., San Francisco, 1987.

39. M. Fleischmann and S. Pons, *Journal of Electroanalytical Chemistry*, **261**, 301 (1989); **263**, 187 (1989).

40. S. E. Jones, E. P. Palmer, J. B. Czirr, D. L. Decker, G. L. Jensen, J. M. Thorne, S. F. Taylor, and J. Rafelski, *Nature*, **338**, 737 (1989).

41. C. Walling and J. Simons, *Journal of Physical Chemistry*, **93**, 4693 (1989).

42. (a) D. E. Williams, D. J. S. Findlay, D. H. Croston, M. R. Sené, M. Bailey, S. Croft, B. W. Hooton, C. P. Jones, A. R. J. Kucernak, K. A. Mason, and R. I. Taylor, *Nature*, **342**, 375 (1989). See also: (b) J. Maddox, *ibid.*, **340**, 15 (1989). (c) D. Swinbanks, *ibid.*, **359**, 765 (1992).

43. The cold fusion episode excited many items in 1989; see *Nature*, Index for Volumes 337-342 (1989). There are also many articles in *Chemical & Engineering News*: (a) R. Dagani, May 22,1989, pp.8-20. (b) November 5,1990, p.4. (c) R. Dagani, January 14,1991, pp.4-5; (d) April 1,1991, pp.31-33. (e) July 1,1991, pp.5-6. (f) January 13,1992, pp.4-5. (g) April 6,1992, p.5; April 13,1992, pp.2-3. (h) R. Dagani, June 14,1993, pp.38-41. (i) R. Dagani, March 20,1995, p.22-23; April 10,1995, pp.7-8.

44. S. Pons and M. Fleischmann, *Physics Letters A*, **176**, 1 (1993).

45. *Wall Street Journal*, July 14,1994, p.1, p.A4.

46. *Fusion Facts,* **4**, 3 (November 1992).

47. (a) M. H. Miles, *Journal of Physical Chemistry*, **98**, 1948 (1994); *Journal of Physical Chemistry B*, **102**, 3642 (1998). (b) S. E. Jones and L. D. Hansen, *Journal of Physical Chemistry*, **99**, 6966 (1995). (c) *Chemical & Engineering News*, June 5,1995, pp.34-35, p.38, p.40. (d) T. O. Passel and G. K. Kohn, *ibid.*, September 11,1995, p.4. (e) S. E. Jones, *ibid.*, December 4,1995, p.5; July 13,1998, pp.10-11. (f) *American Scientist*, **80**, 107 (1992). (g) J. E. Bishop, *Wall Street Journal*, October 19,1992, p.B8.

48. (a) Frank Close, Too Hot to Handle: The Story of the Race for Cold Fusion, Princeton University, Princeton,NJ, 1991. (b) Eugene F. Mallove, Fire from Ice: Searching for the Truth behind the Cold Fusion Furor, John Wiley & Sons, New York, 1991. (c) John R.

Huizenga, Cold Fusion: The Scientific Fiasco of the Century, University of Rochester Press, Rochester,NY, 1992. (d) Gary Taubes, Bad Science: The Short Life and Weird Times of Cold Fusion, Random House, 1993.

49. *New Scientist*, **122**, No.1670, 31 (1989).

50. P. Lowell, Mars and Its Canals, Macmillan Co., New York/London, 1906.

51. (a) T. Vam Flandern, "New Evidence of Artificiality at Cydonia on Mars", Meta Research Bulletin, Vol.6; Internet Web-Site www.metaresearch.org. (b) *The American Spectator*, June 1999, p.76.

52. S. LeVay, *Science*, **253**, 1034 (1991); The Sexual Brain, MIT Press, Cambridge, 1993.

53. D. H. Hamer, S. Hu, V. L. Magnuson, N. Hu, and A. M. L. Pattatucci, *Science*, **261**, 321 (1993); S. Hu, A. M. L. Pattatucci, C. Patterson, L. Li, D. W. Fulker, S. C. Cherry, L. Kruglyak, and D. H. Hamer, *Nature Genetics*, **11**, 248 (1995).

54. *Houston Chronicle*, Houston,TX, July 8,1995, p.4A.

55. (a) M. Eliot, *Science*, **268**, 1841 (1995). (b) C. Thomas, *Human Events*, November 24,1995, p.19. (c) R. Finn, *The Scientist*, January 8,1996, p.13.

56. *Chemical & Engineering News*, December 14,1998, p.80; August 2,1999, p.64.

57. L. Rosa, E. Rosa, L. Sarner, and S. Barrett, *JAMA* (*Journal of the American Medical Association*), **279**, 1005 (1998).

58. D. W. Forest, Francis Galton: The Life and Work of a Victorian Genius, Taplinger Publishing Co., New York, 1974, pp.110-113.

59. (a) L. R. Ember, *Chemical & Engineering News*, December 7,1998, p.14. (b) S. Bunk, *The Scientist*, February 1,1999, p.10.

60. (a) F. Haber and R. Willstätter, *Berichte der Deutsche Chemische Gesellschaft*, **64**, 2844 (1931); (b) F. Haber and J. Weiss, *Proceedings of the Royal Society of London*, [A], **147**, 332 (1934).

61. (a) J. H. Ludens, M. A. Clark, D. W. DuCharme, D. W. Harris, B. S. Lutzke, F. Mandel, W. R. Matthews, D. M. Sutter, and J. M. Hamlyn, *Hypertension*, **17**, 923 (1991). (b) W. R. Matthews, D. W. DuCharme, J. M. Hamlyn, D. W. Harris, F. Mandel, M. A. Clark, and J. H. Ludens, *Hypertension*, **17**, 930 (1991). (c) J. M. Hamlyn, M. P. Blaustein, S. Bova, D. W. DuCharme, D. W. Harris, F. Mandel, W. R. Matthews, and J. H. Ludens, *Proceedings of the National Academy of Sciences USA*, **88**, 6259 (1991). (d) J. Laredo, B. P. Hamilton, and J. M. Hamlyn, *Endocrinology*, **135**, 794 (1994). (e) B. P. Hamilton, P. Manunta, J. Laredo, J. H. Hamilton, and J. M. Hamlyn, *Current Opinion in Endocrinology and Diabetes*, **1**, 123 (1994). (f) W. Schoner, *Progress in Drug Research*, **41**, 249 (1993).

62. A. Kawamura, J. Guo, Y. Hagaki, C. Bell, Y. Wang, G. T. Hauport,

S. Magil, R. T. Gallagher, N. Berova, and K. Nakanishi, *Proceedings of the National Academy of Sciences USA*, **96**, 6654 (1999).

63. (a) M. Waldholz, *Wall Street Journal*, March 16,1995, p.B1; February 5,2001, p.A20. (b) R. T. King, *ibid.*, November 12,1998, p.A1; February 10,1999, p.B1. (c) R. Winslow, *ibid.*, April 6,1999, p.B1. (d) S. Brownlee, *U.S. News & World Reports*, February 22,1999, p.63. (e) P. Smaglik, *The Scientist*, March 1,1999, p.1. (f) T. Browder, J. Folkman, and S. Pirie-Shepherd, *Journal of Biological Chemistry*, **275**, 1521 (2000). (g) R. Cooke, Dr. Folkman's War, Random House, New York, 2001.

64. (a) R. A. B. Ezekowitz, J. B. Mulliken, and J. Folkman, *New England Journal of Medicine*, **326**, 1456 (1992); A. Ezekowitz, J. Mulliken, and J. Folkman, *ibid.*, **330**, 300 (1994); **333**, 595 (1995). (b) R. Dalton, *Science*, **377**, 569 (1995).

65. (a) *Houston Chronicle*, Houston,TX, May 27,1997. (b) *U.S. News & World Reports*, October 5,1998, pp.28-35.

66. G. Poinar, *Experientia*, **50**, 536 (1994); *American Scientist*, **87**, 446, 486 (1999).

67. (a) T. Lindahl, *Cell*, **90**, 1 (1997). (b) J. J. Austin, A. B. Smith, and R. H. Thomas, *Trends in Ecology and Evolution*, **12**, 303 (1997). (c) K. K. O. Walden and H. M. Robertson, *Molecular Biology and Evolution*, **14**, 1075 (1997). (d) H. M. Robertson, *American Scientist*, **87**, 485 (1999).

68. G. Gutiérrez and A. Marín, *Molecular Biology and Evolution*, **15**, 926 (1998).

69. D. S. McKay, E. K. Gibson, K. L. Thomas-Keprta, H. Vali, C. S. Romanek, S. J. Clemett, X. D. F. Chillier, C. R. Maechling, and R. N. Zare, *Science*, **273**, 924 (1996).

70. (a) J. P. Bradley, R. P. Harvey, and H. Y. McSween, *Geochimica et Cosmochimica Acta*, **60**, 5149 (1996). (b) E. Wilson, *Chemical & Engineering News*, January 6,1997, p.9.

71. S. G. Philander, Is the Temperature Rising? The Uncertain Science of Global Warming, Princeton University Press, Princeton,NJ, 1998.

CHAPTER 6. ERROR, HOAX, MISCONDUCT, FRAUD, & ETHICS

1. J. R. Sabine, "The Error Rate in Biological Publication: A Preliminary Survey", Science, Technology, & Human Values, John Wiley & Sons, **10**, [No.1], 62-69 (1985).

2. H. Zuckerman, "Deviant Behavior and Social Control in Science", in E. Sagarin, Editor, Deviance and Social Change, Sage, Beverly Hills,CA, 1977, pp.87-138.

3. (a) *Chemical & Engineering News*, November 23,1998, p.37. (b) E. Wilson, *ibid.*, December 17,2001, p.11.

4. (a) J. Gumulka and L. L. Smith, *Journal of the American Chemical Society*, **105**, 1972 (1983). (b) K. Jaworski and L. L. Smith, *Journal of Organic Chemistry*, **53**, 545 (1988). (c) K. Jaworski and L. L. Smith, *Magnetic Resonance in Chemistry*, **26**, 104 (1988). (d) L. L. Smith, E. L. Ezell, and K. Jaworski, *Steroids*, **61**, 627 (1996).

5. (a) C. Blinderman, The Piltdown Inquest, Prometheus Books, Buffalo,NY, 1986. (b) F. Spencer, Piltdown: A Scientific Forgery, Oxford University Press, London/New York, 1990. (c) J. E. Walsh, Unraveling Piltdown: The Science Fraud of the Century and Its Solution, Random House, New York, 1996.

6. A. D. Sokal, *Social Text*, **46-47**, 217 (1996).

7. (a) A. Sokal, *Lingua Franca*, **6**[4], 62 (1996). (b) A. Sokal and J. Bricmont, Fashionable Nonsense, St. Martin's Press, New York, 1998. See also (c) I. M. Klotz, *The Scientist*, July 22,1996, p.9; (d) S. Weinberg, *New York Review of Books*, August 8,1996, p.11; (e) P. R. Gross, *The Scientist*, April 28,1997. (f) N. Koertge, A House Built on Sand: Exposing Postmodernist Myths About Science, Oxford University Press, Oxford, 1998. (g) *Lingua Franca*, The Sokal Hoax: The Sham That Shook the Academy, University of Nebraska Press, Lincoln, 2000.

8. P. Zurer, *Chemical & Engineering News*, July 4,1994, p.20.

9. K. D. Hansen and B. C. Hansen, *FASEB Journal*, **5**, 2512 (1991).

10. (a) Committee on the Responsible Conduct of Research, Institute of Medicine, The Responsible Conduct of Research in the Health Sciences, National Academy Press, Washington City,DC, 1989. (b) "On Being a Scientist", *Proceedings of the National Academy of Sciences USA*, **86**, 9058 (1989). (c) Panel on Scientific Responsibility and the Conduct of Research, National Academy of Sciences/National Academy of Engineering/Institute of Medicine, Responsible Science: Ensuring the Integrity of the Research Process, National Academy Press, Washington City,DC, 1992. (d) Committee on the Conduct of Science, National Academy of Sciences, On Being a Scientist, National Academy Press, Washington City,DC, 1989; Committee on Science, Engineering, and Public Policy, National Academy of Sciences, On Being a Scientist. Responsible Conduct in Research, National Academy Press, Washington City,DC, 1995. (e) J. Woodward and D. Goodstein, *American Scientist*, **84**, 479 (1996).

11. *NIH Guide for Grants and Contracts*, **11**, No.8, 1 (1982); **15**, No.11 (1986).

12. J. Long, *Chemical & Engineering News*, January 23,1989, p.61.

13. (a) P. Zurer, *Chemical & Engineering News*, September 26,1988, p.6; March 23,1992, p.14; March 30,1992, p.5. (9) P. Gigot, *Wall Street Journal*, August 16,1991, p.A8.

14. P. S. Zurer, *Chemical & Engineering News*, February 3,1992, p.17.

15. *NIH Guide for Grants and Contracts*, **23**, No.30, 2 (1994).
16. P. S. Zurer, *Chemical & Engineering News*, August 28,1989, p.22.
17. P. S. Zurer, *Chemical & Engineering News*, July 3,1995, p.14; *FASEB Newsletter*, **29**, No.4, p.1 (1996).
18. (a) *Federal Register*, **64**, 55722 (1999); (b) B. Hileman, *Chemical & Engineering News*, October 25,1999, p.12; W. Schulz, *ibid.*, November 29,1999, p.28.
19. H. K. Schachman, *Science*, **261**, 148 (1993).
20. K. Kaneyuki and K. Scholberg, *American Scientist*, **87**, 222 (1999).
21. (a) *ASBMB NEWS*, **9**, No.4, 4 (2000). (b) *Chemical & Engineering News*, December 11,2000, p.37.
22. *The Scientist*, March 31,1997, p.1, p.3; April 15,1997, p.31.
23. (a) P. Zurer, *Chemical & Engineering News*, April 18,1994. (b) *Houston Post*, Houston,TX, April 14,1994; May 19,1994. (c) E. Garfield, *The Scientist*, November 27,1995.
24. *The Scientist*, June 9,1997, p.1, p.3.
25. D. Weaver, M. H. Reis, C. Albanese, F. Constantini, D. Baltimore, and T. Imanishi-Kari, *Cell*, **45**, 247 (1986).
26. T. Imanishi-Kari, D. Weaver, and D. Baltimore, *Cell*, **57**, 515 (1989).
27. D. Weaver, C. Albanese, F. Constantini, and D. Baltimore, *Cell*, **64**, 536 (1991).
28. T. Imanishi-Kari, C. A. Huang, J. Iacomini, and N. Yannoutsos, *Journal of Immunology*, **150**, 3311, 3327 (1993).
29. (a) Judy Sarasohn, Science On Trial: The Whistle-Blower, the Accused, and the Nobel Laureate, St. Martin's, 1993. (b) D. J. Kevles, *New Yorker*, May 27,1996. (c) *Wall Street Journal*, June 24,1996, p.B8; July 2,1996, p.A12. (d) P. Zurer, *Chemical & Engineering News*, February 20,1989, p.5; June 24, 1996, p.31. (e) R. Rawls, *ibid.*, July 1,1996, p.6. (e) D. J. Kevles, The Baltimore Case: A Trial of Politics, Science, and Character, W. W. Norton & Co., New York/London, 1998. (f) David L. Hull, *New York Review of Books*, **45**, No.19, 24-30 (1998). (g) S. Crotty, Ahead of the Curve: David Baltimore's Life in Science, University of California Press, Berkeley/Los Angeles, 2001.
30. D. Dickson, *Nature*, **370**, 315 (1994).
31. M. E. Watanabe, *The Scientist*, February 5,1997, p.1.
32. (a) M. B. Brennan, *Chemical & Engineering News*, January 25,1999, p.11. (b) J. D. Altom, *ibid.*, July 26,1999, p.39.
33. P. S. Bernstein, W. C. Law, and R. R. Rando, *Proceedings of the National Academy of Sciences USA*, **84**, 1849 (1987).
34. C. D. Bridges and R. A. Alvarez, *Science*, **236**, 1678 (1987).
35. *Chemical & Engineering News*, June 29,1987, p.22.
36. (a) D. Nok, *Chemical & Engineering News*, September 14,1987, p.51. (b) L. Birladeanu, *Chemical & Engineering News*, September

21,1987, p.5.

37. (a) B. J. Culliton, *Science*, **245**, 120 (1989). (b) P. S. Zurer, *Chemical & Engineering News*, August 7,1989, pp.24-25.

38. (a) L. A. Paquette, M. A. Kesselmayer, G. E. Underiner, S. D. House, R. D. Rogers, K. Meerholz, and J. Heinze, *Journal of the American Chemical Society*, **114**, 2644 (1992). (b) *Federal Register*, June 21,1993, p.33830. (c) *NIH Guide for Grants and Contracts*, **22**, No.23, 2 (1993). (d) *Chemical & Engineering News*, July 12,1993, pp.22-34; August 16,1993, p.5; March 21,1994, p.18.

39. G. Putka, *Wall Street Journal*, July 12,1991, p.1.

40. A. Grafton, Forgers and Critics: Creativity and Duplicity in Western Scholarship, Princeton University Press, New Haven, 1990.

41. *Chemical & Engineering News*, November 4,1996, p.7; *The Scientist*, November 11,1996, p.30. See particularly (a) A. Hajra, P. P. Liu, Q. Wang, C. A. Kelly, T. Stacy, R. S. Adelstein, N. A. Speck, and F. S. Collins, *Proceedings of the National Academy of Sciences USA*, **92**, 1926 (1995); R. S. Adelstein, F. S. Collins, A. Hajra, C. A. Kelly, P. P. Liu, N. A. Speck, T. Stacy, and Q. Wang, *ibid.*, **93**, 15523 (1996). (b) C. Wijmenga, P. E. Gregory, A. Hajra, E. Schrock, T. Ried, R. Eils, P. P. Liu, and F. S. Collins, *ibid.*, **93**, 1630, 15522 (1996).

42. (a) B. Culliton, *Science*, May 10,1975; June 14,1974. (b) J. R. Hixon, The Patchwork Mouse, Anchor Press, Garden City,NY, 1976. (c) *Proceedings of the National Academy of Sciences USA*, **86**, 9068 (1989).

43. (a) A. Bienkowski, "Fudging in Science--A Review", *The Bookman* (UTMB), **5**, No.6, June 1978. (b) Ellen S. Moore, "Painting the Mice", *Medical Humanities Rounds* (UTMB), **9**, No.5, January 1992; *The Chronicle. The Newsletter of the Institute for the Medical Humanities* (UTMB), **10**, No.2, Fall 1992.

44. (a) R. Knox, *JAMA*, **249**, 1797 (1983). (b) B. J. Culliton, *Science*, **219**, 937 (1983); **220**, 936 (1983). (c) C. Wallis, *Time*, February 28,1983, p.49.

45. W. W. Stewart and N. Feder, *Nature*, **325**, 207 (1987).

46. *Nature*, **325**, 181 (1987).

47. E. Braunwald, *Nature*, **325**, 215 (1987).

48. (a) C. J. Glueck, M. J. Mellies, M. Dine, T. Perry, and P. Laskarzewski, *Pediatrics*, **78**, 338 (1986); (b) Retraction: C. J. Glueck, P. Laskarzewski, M. J. Mellies, and T. Perry, *ibid.*, **80**, 766 (1987).

49. (a) W. J. Broad, *Science*, **212**, 1367 (1981); **216**, 1081 (1982); (b) M. Sun, *Science*, **212**, 1366 (1981); **219**, 270 (1983). (c) D. Dickson, *Nature*, **295**, 543 (1982).

50. (a) N. Wade, *Science*, **194**, 916 (1976); (b) O. Gillie, *ibid.*, **204**, 1035 (1979); (c) L. J. Cronbach, *ibid.*, **206**, 1392 (1979). (d) L. S.

Hearnshaw, Cyril Burt, Psychologist, Cornell University Press, Ithaca,NY, 1979.

51. (a) D. Brand, *Time*, June 1,1987, p.59. (b) C. Holden, *Science*, **235**, 1566 (1987). (c) P. Zurer, *Chemical & Engineering News*, April 25,1988, p.5; September 26,1988, p.6. (d) E. Garfield and A. Welljams-Dorof, *JAMA*, **263**, 1424 (1990).
52. B. Morris, *Wall Street Journal*, March 1,1983, p.12.
53. D. Hanson, *Chemical & Engineering News*, June 29,1987, p.19; W. Bogdanich, *Wall Street Journal*, **80**, No.23, p.1 (1987).
54. (a) C. G. Sibley, J. A. Comstock, and J. E. Ahlquist, *Journal of Molecular Evolution*, **30**, 202 (1990). (b) J. Marks, *American Scientist*, **81**, 380 (1993).
55. (a) *Medical Research Funding*, September 15,1990, p.6. (b) P. Zurer, *Chemical & Engineering News*, August 20,1990, p.6. (c) G. Taubes, *Science*, **263**, 605 (1994).
56. (a) S. Kawabata, G. A. Higgins, and J. W. Gordon, *Nature*, **354**, 476 (1991). (b) *Houston Post*, Houston,TX, February 29,1992, p.A-13.
57. (a) R. Locke, *Nature*, **324**, 401 (1986). (b) R. Dalton, *The Scientist*, May 18,1987, p.1, p.8.
58. *Wall Street Journal*, January 27,1993.
59. F. A. Cotton, *The Chemical Intelligencer*, **5**, No.1, 43 (1999).
60. (a) F. Badger, *Mathematics Magazine*, **67**, 83 (1994). (b) J. Maddox, *Nature*, **370**, 323 (1994).
61. (a) G. S. Hammond and A. Ravve, *Journal of the American Chemical Society*, **73**, 1891 (1951). (b) G. S. Hammond, *The Spectrum*, **7**, No.1, p.10 (1994). (c) R. A. Benkeser, R. B. Gosnell, and W. Schroeder, *Journal of the American Chemical Society*, **79**, 2339 (1957).
62. R. Lewin, *Science*, **244**, 277 (1989).
63. *The Scientist*, July 20,1998, p.1, pp.5-6.
64. *Chemical & Engineering News*, October 24,1994, p.9.
65. (a) J. Anderson, *The Scientist*, July 10,1989, p.1; October 16,1989, p.8. (b) R. S. Desowitz, The Malaria Capers: More Tales of Parasites and People, Research and Reality, W. W. Norton & Co., New York, 1991.
66. W. J. Broad, *Science*, **207**, 743 (1980).
67. (a) Ethical Guidelines to Publication of Chemical Research, *Chemical & Engineering News*, September 26,1983, pp.39-43. (b) American Chemical Society, *Accounts of Chemical Research*, **27**, 179 (1994).
68. *NIH Guide for Grants and Contracts*, **19**, No.30, 1 (1990).
69. *American Medical News*, October 20,1989, p.1, p.42.
70. (a) R. L. Penslar, Editor, Research Ethics: Cases and Materials, Indiana University Press, Indianapolis,IN, 1995. (b) E. Erwin, S. Gendin, and L. Kleiman, Editors, Ethical Issues in Scientific

Research: An Anthology, Garland Publishing Inc., 1994. (c) K. Shrader-Frechette, Ethics of Scientific Research, Rowman & Littlefield Publishers Inc., 1994.
71. M. Mohr, *International Science Reviews*, **4**, 45 (1979).

CHAPTER 7. SOME OTHER MATTERS

1. P. Duesberg, *Science*, **241**, 514 (1988); *New York Review of Books*, May 23,1996, p.14; August 8,1996, p.51.
2. (a) P. H. Duesberg, Inventing the AIDS Virus, Regnery Publishing Inc., Washington City,DC, 1997. (b) P. H. Duesberg, Infectious AIDS: Have We Been Misled?, North Atlantic Books, Berkeley,CA, 1995. (c) P. H. Duesberg, Editor, AIDS: Virus- or Drug Induced?, Kluwer Academic Press, 1996. (d) B. J. Ellison and P. Duesberg, Why We Will Never Win the War on AIDS, Inside Story Communications, El Cerrito,CA, 1994.
3. (a) J. Maddox, *Nature*, **363**, 109 (1993). (b) J. Moore, *ibid.*, **380**, 293 (1996).
4. R. W. Smith, *The Scientist*, January 22, 2001, p.39.
5. (a) B. A. Palevitz and R. Lewis, *The Scientist*, December 8,1997, p.8. (b) R. Lewis, *ibid.*, May 10,1999, p.1, p.8.
6. T. L. James, *Proceedings of the National Academy of Sciences USA*, **94**, 10086 (1997).
7. The 1994/1995 Calbiochem Biochemicals Catalog, Calbiochem-Novachem Corp., San Diego,CA, p.47.
8. B. Hileman, *Chemical & Engineering News*, October 26,1998, p.31, pp.34-35.
9. B. D. Davis, *The Scientist*, February 20,1989, p.9.
10. (a) L. R. Raber, *Chemical & Engineering News*, December 21,1998, pp.24-25; April 12,1999, p.11. (b) W. G. Schulz, *ibid.*, March 22,1999, p.23. (c) *The Scientist*, February 15,1999, p.1, p.8. (c) *Federal Register*, February 4,1999, p.5684. (d) L. McGinley, *Wall Street Journal*, March 1,1999, p.A24; June 7,1999, p.A22. (e) J. Couzin, *U.S. News & World Reports*, March 29,1999, p.70.
11. R. T. King, *Wall Street Journal*, February 2,1999, p.B1.
12. W. Schulz, *Chemical & Engineering News*, December 18,2000, p.5.
13. (a) E. O. Laumann, A. Pauk, and R. C. Rosen, *JAMA*, **281**, 537 (1999); (b) Associated Press, *Austin American-Statesman*, Austin,TX, February 11,1999, p.A5.
14. R. Damadian, *Science*, **171**, 1151 (1971).
15. D. P. Hollis, L. A. Saryan, and H. P. Morris, *Johns Hopkins Medical Journal*, **131**, 441 (1972).
16. A. Schatz and S. Waksman, *Proceedings of the Society for Experimental Biology and Medicine*, **57**, 244 (1944).
17. K. S. Zaner and R. Damadian, *Science*, **189**, 729 (1975); K. S. Zaner

and R. Damadian, *Physiological Chemistry & Physics*, **9**, 473 (1977).

18. K. S. Zaner and R. Damadian, *Physiological Chemistry & Physics*, **7**, 437 (1975).

19. J. A. Koutcher and R. Damadian, *Physiological Chemistry & Physics*, **9**, 181 (1977).

20. D. P. Hollis, Abusing Cancer Science. The Truth about NMR and Cancer, Strawberry Fields Press, Chehalis,WA, 1987.

21. E. T. Fossel, J. M. Carr, and J. McDonagh, *New England Journal of Medicine*, **315**, 1369 (1986).

22. P. S. Schein, *New England Journal of Medicine*, **315**, 1410 (1986).

23. M. Clark and S. Doherty, *Newsweek*, December 8,1986, p.86.

24. A series of seven articles "Detection of Malignant Tumors by Nuclear Magnetic Resonance Spectroscopy of Plasma", *New England Journal of Medicine*, **316**, 1411-1415 (1987).

25. P. Zurer, *Chemical & Engineering News*, December 21,1987, p.6.

26. (a) P. Wilding, M. B. Senior, T. Inubushi, and M. L. Ludwick, *Clinical Chemistry*, **34**, 505 (1988). (b) S. Tran Dinh, M. Schlumberger, C. Gicquel, J. M. Neumann, M. Herve, G. Turpin, and C. Parmentier, *Bulletin du Cancer*, **75**, 795 (1988).

27. (a) T. Engan, J. Krane, O. Klepp, and S. Kvinnsland, *New England Journal of Medicine*, **322**, 949 (1990). (b) P. Okunieff, A. Zietman, J. Kahn, S. Singer, L. J. Neuringer, R. A. Levine, and F. E. Evans, *ibid.*, **322**, 953 (1990). (c) R. Shulman, *ibid.*, **322**, 1002 (1990). (d) *ibid.*, **323**, 679 (1990).

28. (a) E. T. Fossel, J. M. Clark, and J. McDonagh, *New England Journal of Medicine*, **316**, 1415 (1987). (b) E. T. Fossel, *ibid.*, **323**, 677 (1990).

29. (a) A. E. Wright, *Medical Physics*, **14**, 473 (1988). (b) A. E. Wright, "^{1}H Characterization of Human Blood Plasma", World Congress on Medical Physics, San Antonio,TX, August 6-12,1988.

30. (a) R. Dagani, *Chemical & Engineering News*, April 29,1996, p.69; November 18,1996, p.9. (b) E. T. Mallove, *ibid.*, June 17,1996, p.4.

31. D. Schneider, *American Scientist*, **87**, 314 (1999).

32. D. A. Price, *American Spectator*, January 1994, p.31.

33. M. Finkin, The Case for Tenure, ILR Press, Ithaca,NY, 1996.

34. R. M. O'Neill, *Academic Physician Scientist*, January 1997, pp.9-11.

35. *Wall Street Journal*, November 13,1996, p.A23.

36. (a) *TFA (Texas Faculty Association) Bulletin*, **4**, No.1, p.1, p.3 (1988). (b) The Vanguard, The Official Newsletter of the Galveston Medical Branch Chapter of the Texas Faculty Association, **1**, No.1 (1996).

37. (a) *TFA (Texas Faculty Association) Bulletin*, **11**, No.1, August/September 1996; **11**, No.2, February 1997. (b) *Austin American-Statesman*, Austin,TX, Nov.15,1996.

38. *Austin American-Statesman*, Austin,TX, February 26,1997, p.B2.

39. (a) L. L. Smith and E. L. Ezell, *Steroids*, **53**, 513 (1989). (b) E. J. Delany, R. G. Sherrill, V. Palaniswamy, T. C. Sedergran, and S. P. Taylor, *ibid.*, **59**, 196 (1994).

40. (a) J. E. van Lier and L. L. Smith, *Biochemistry*, **6**, 3269 (1967). (b) J. E. van Lier and L. L. Smith, *Texas Reports on Biology & Medicine*, **27**, 167 (1969). (c) L. L. Smith and J. E. van Lier, *Atherosclerosis*, **12**, 1 (1970); **13** 140 (1971). (d) J. E. van Lier and L. L. Smith, *Lipids*, **6**, 85 (1971). (e) L. L. Smith, D. R. Ray, J. A. Moody, J. D. Wells, and J. E. van Lier, *Journal of Neurochemistry*, **19**, 889 (1972). (f) Y. Y. Lin and L. L. Smith, *Biochimica et Biophysica Acta*, **348**, 189 (1974). (g) Y. Y. Lin and L. L. Smith, *Journal of Neurochemistry*, **25**, 659 (1975). (h) J. I. Teng and L. L. Smith, *Texas Reports on Biology & Medicine*, **33**, 293 (1975). (i) L. L. Smith, J. I. Teng, Y. Y. Lin, P. K. Seitz, and M. F. McGehee, *Journal of Steroid Biochemistry*, **14**, 889 (1981).

41. (a) G. Steel, C. J. W. Brooks, and W. A. Harland, *Biochemical Journal*, **99**, 51P (1966). (b) C. J. W. Brooks, W. A. Harland, and G. Steel, *Biochimica et Biophysica Acta*, **125**, 620 (1966). (c) R. Fumagalli, G. Galli, and G. Urna, *Life Sciences*, Part 2, **10**, 25 (1971).

42. (a) K. L. H. Carpenter, S. E. Taylor, J. A. Ballentine, B. Fussell, B. Halliwell, and M. J. Mitchinson, *Biochimica et Biophysica Acta*, **1167**, 121 (1993). (b) K. L. H. Carpenter, S. E. Taylor, C. van der Veen, B. K. Williamson, J. A. Ballentine, and M. J. Mitchison, *ibid.*, **1256**, 141 (1995).

43. (a) I. Ginsburg, "The Disregard Syndrome: A Menace to Honest Science", *The Scientist*, December 10,2001. (b) E. Garfield, "Demand Citation Vigilance", *ibid.*, January 21,2002, p.6, pp.12-13. (c) C. G. Daughton, "Literature Forensics: Navigating Through Flotsam, Jetsam, and Lagan", *ibid.*, February 18,2002, p.12.

44. D. B. Kates, H. E. Schaffer, J. K. Lattimer, G. B. Murray, and E. H. Cassem, *Tennessee Law Review*, **62**, 513 (1995).

45. G. D. Lundberg and A. Flanagan, *JAMA*, **262**, 2003 (1989). (b) Bill Silberg, *American Medical News*, October 20,1989, p.19.

46. N. Düzgünes, *The Scientist*, April 12,1999, p.13.

47. (a) S. Cole, J. R. Cole, and G. A. Simon, *Science*, **214**, 881 (1981). (b) A. E. Sowers, *The Scientist*, October 16,1995, p.12.

48. W. Schultz, *Chemical & Engineering News*, October 11,1999, p.72.

49. S. Oates, With Malice Towards None: The Life of Abraham Lincoln, Harper & Row, New York, 1977.

50. (a) *Chemical & Engineering News*, April 19,1993, p.28; May 24,1993, p.24; June 21,1993, p.8; September 27,1993, p.21; October 4,1993, p.16. (b) *The Scientist*, May 17,1993, p.1, p.4, p.5, p.7; May 17,1993, p.11, p.15; June 28,1993, p.4; November 1,1993, p.1, p.4,

p.20, p.22. March 20,1995, p.30. (c) *Wall Street Journal*, April 26,1993.

51. (a) W. W. Stewart and N. Feder, *The Scientist*, December 14,1987, p.13. (b) P. Zurer, *Chemical & Engineering News*, October 15,1990, p.8; January 11,1993, p.4; June 14,1993, p.6; November 15,1993, p.8; January 9,1995, p.5. (c) *Wall Street Journal*, December 31,1992, p.A3. (d) P. Kefalides, *The Scientist*, April 3,1995, p.1.

52. M. P. Glazer and P. M. Glazer, The Whistleblowers, Basic Books, New York, 1989.

53. B. Goodman, *The Scientist*, August 18,1997, p.1.

54. R. L. Sprague, B. W. Hollis, and J. Jacobstein, *The Scientist*, December 14,1987, p.11.

55. J. J. Fialka, *Wall Street Journal*, June 18,1999, p.A22.

56. (a) F. Hoke, *The Scientist*, May 15,1995, p.1; September 4,1995, p.1. (b) T. Devine, *ibid.*, May 15,1995, p.11. (c) T. W. Durso, *ibid.*, February 19,1996, p.3. (d) B. Goodman, *ibid.*, January 22,1996, p.1; July 22,1996, p.3. (e) D. L. Burk, *ibid.*, September 16,1996, p.9.

INDEX

Alphabetized index entries are for items encountered in the text, by page numbers. Items in the References section are not indexed.

Lightning Source UK Ltd.
Milton Keynes UK
UKOW06f0848240816

281304UK00001B/105/P